Climate Change and Hydrosphere

The Water Planet in Crises

Climate Change and Hydrosphere

The Water Planet in Crises

— *Editors* —

Vir Singh

Bipin Kumar

Nanda Nautiyal

Akanksha Rastogi

Anil Kumar Shankhwar

BIOTECH

2012

BIOTECH BOOKS®

ISBN 978-81-7622-257-0

Published by: **BIOTECH BOOKS®**
4762-63/23, Ansari Road, Darya Ganj,
New Delhi - 110 002
Phone: +91-011-23262132
E-mail: biotechbooks@yahoo.co.in

Printed at: **Chawla Offset Printers**
Delhi - 110 052

PRINTED IN INDIA

Preface

Our planet is a water planet. Water, that is, basis of life on Earth! But this very basis of life is in crises. Human behaviour with water has been erroneous. Advances in science and technology have further contributed to intensify water problems. Development in agriculture culminating into pesticide-dependent food production practices has poisoned potable water sources – surface water as well as groundwater – virtually everywhere on Earth! Pesticides, via water, have linked with virtually all food chains and complex food web. Very few water sources have been left which are not contaminated. Contamination of water with pesticides, heavy metals, artificial dyes, deadly chemicals as well as with all sorts of microorganism appears to have become a rule of nature thanks to extensive anthropogenic intervention leading to infestation of all components of the biosphere.

The second factor to affect water resources is a recorded change in climate pattern of the globe. A climate change is not an isolated phenomenon. This is to affect all components and all processes of biosphere. Water pollution largely due to anthropogenic intervention coupled with climate change, which is also supposedly owing to anthropogenic dimensions, become the major cause of water crises on the water planet.

There appears to be no possibility of reverting the trend of increasing water crises in near future. The global climate crises are becoming water crises and vice versa. Pollutants in atmosphere is linked with water pollution. Pollution in soil also becomes water pollution. It is water which is available in all the components of biosphere. It is the water which exists in all the three forms of matter. It is the water which, if spoiled, the whole life is bound to be affected. If water is itself ill, the whole life is bound to be in illness. As the water crises, including potable water scarcity and water pollution, are intensifying miseries of life are aggravating.

The other issue is that of excessive volume of water experienced during a short period which is thanks to deforestation, climate change leading to disrupted weather cycle, incessant rains over a short period, and mismanagement leading to failure in controlling floods. Building up huge reservoirs by damming large rivers is no way to control floods. Besides, large dams increase risks to life and property. Large inundated areas due to high dams also become a reason to aggravate environmental, economic and often seismological problems. Building up high dams in sensitive zones like geologically instable Himalayas is no flawless as it faces numerous risks to pristine ecosystems in upland and lowland areas.

Accelerated melting of glaciers due to global warming is another issue facing contemporary times. It should be reminded that impact of global warming in the Himalayan mountains is more severe than in the plains. Excessive rate of melt is a conspicuous indicator of the impact of global warming in the Himalayan mountains. If this trend continues unabatedly for longer duration, the entire South Asian region would experience acute shortage of drinking water in future, particularly in lean period. Further, accelerated melt of glaciers would lead to swelling of seas and oceans. A few meter increase in sea level would turn catastrophic for the inhabitants in the coastal areas.

Himalayas are home to snow. They are also often referred to as "The Third Pole" as most of the snow and ice outside two Poles of the Earth – the North Pole and the South Pole – is deposited in the Himalayas. A state of global warming and climate change threatens the Third Pole.

The volume in your hands is devoted to water and water planet. It focuses on water crises borne out both by human-created pollution and climate change. Potable water resources are shrinking. Hydrological cycle associated with forests is breaking down thanks to clearing of forest areas. Rainfed rivers are becoming seasonal. Whatever little water is available for drinking purpose is polluted.

So what should be done? Where to go from here? How could we restore water cycle? We know the answer to these questions, but we are too paralysed to resort to any concrete action. Political will power is totally lacking. We cannot afford to dwell in a world being turned into an uninhabitable place for the coming up generations.

Cooling of the Earth is a waking call of our times. Cooling of the Earth would be possible through avoiding and controlling environmental pollution by stopping or minimising carbon emissions and developing and adopting renewable energy sources, conserving natural resources, such as natural forests, water resources, and through enhanced carbon sequestration by augmentation of natural resources.

Divided into three parts the chapters in the book focus on water crises due to pollution, climate change impact on hydrosphere and on various response to water crises including case studies. The data, views and interpretations expressed are those of the individual writers and do not necessarily reflect the opinion of the editors, their organization and publisher.

We hope the volume being presented would be useful in providing information necessary to contribute to creating awareness about our water resources vital for the sustenance of life on Earth. We hope the book will also help provide crucial information about the appropriate management of our resources and environment.

Vir Singh

Bipin Kumar

Nanda Nautiyal

Akanksha Rastogi

Anil Kumar Shankhwar

Contents

Part II
WATER POLLUTION: A THREAT TO LIFE

Part III
CLIMATE CHANGE AND WATER RESOURCES

Part IV
RESPONSE TO WATER CRISES

List of Contributors

Arora, I.M. Rawat Bhawan Awagarh Compound, Mallital, Nainital
E-mail: indry@in.com

Bahukhandi, Kanchan Deoli. University of Petroleum and Energy Studies, Dehradun, Uttarakhand

Baliyan, Deeksha. College of Biotechnology, Sardar Vallabhbhai Patel University of Agriculture and Technology, Meerut
E-mail: deeksha.baliyan@gmail.com

Bhatnagar, SK. College of Biotechnology, Sardar Vallabhbhai Patel University of Agriculture and Technology, Meerut

Chaddha, B. G.B. Pant University of Agriculture & Technology, Pantnagar – 263 145, U.S. Nagar, Uttarakhand
E-mail: bhanukbc6@gmail.com

Dikshit, A.K. Division of Agricultural Chemicals, Indian Agricultural Research Institute, New Delhi – 110 012
E-mail: anandkd_kk49@yahoo.co.in

Gupta, S.C. University of Petroleum and Energy Studies, Dehradun, Uttarakhand

Jose, P.G. School of Environmental Sciences, Jawaharlal Nehru University, New Delhi

Kaphaliya, B. Department of Environmental Sciences, College of Basic Science, G.B. Pant University of Agriculture & Technology, Pantnagar – 263 145, U.S. Nagar, Uttarakhand
E-mail: bhumikafalia@gmail.com

Kaushik, N. Department of Geography, University of Rajasthan, Jaipur
E-mail: nidhikaushik07@yahoo.com

Khanam, Zeba. Department of Environmental Science, College of Basic Sciences and Humanities, G.B. Pant University of Agriculture & Technology, Pantnagar – 263 145, U.S. Nagar, Uttarakhand

Kumar, B. Department of Environmental Science, College of Basic Sciences and Humanities, G.B. Pant University of Agriculture & Technology, Pantnagar – 263 145, U.S. Nagar, Uttarakhand
E-mail: envbipin@gmail.com

Kumar, Neeraj. I.E.T., Alwar
E-mail: neee.kr@gmail.com

Kumar, R. Department of Biotechnology, G.B. Pant University of Agriculture & Technology, Pantnagar – 263 145, U.S. Nagar, Uttarakhand
E-mail: professor.agch@gmail.com

Linda, A. School of Environmental Sciences, Jawaharlal Nehru University, New Delhi

Mahra, Girijesh Singh. M.Sc. Extension Education (Junior Research Fellow), (I.AG.Sc.) B.H.U., Varanasi – 221 005

Mandappa, Pawan. University of Petroleum and Energy Studies, Dehradun, Uttarakhand

Matta, V. College of Technology, G.B. Pant University of Agriculture & Technology, Pantnagar – 263 145, U.S. Nagar, Uttarakhand
E-mail: mattavaibhav@gmail.com

Mishra, C. Animal Genetics Division, Indian Veterinary Research Institute, Izatnagar – 243 122
E-mail: drchinmoymishra@gmail.com

Mishra, K. Department of Environmental Sciences, H.N.B. Garhwal University, Srinagar – Garhwal, Uttarakhand

Mujumdar, Sanskriti S. Reader, Department of Civil Engineering, Faculty of Technology & Engineering, The Maharaja Sayajirao University of Baroda, Vadodara, Gujarat

Nautiyal, Nanda. Department of Environmental Science, College of Basic Sciences and Humanities, G.B. Pant University of Agriculture & Technology, Pantnagar – 263 145, U.S. Nagar, Uttarakhand
E-mail: nautynan@gmail.com

Pant, Aali. College of Technology, G.B. Pant University of Agriculture & Technology, Pantnagar – 263 145, U.S. Nagar, Uttarakhand

Patel, A.S. The M.S. University of Baroda, Vadodara, Gujarat

Patel, Arvind S. Professor & Head, Department of Civil Engineering, Faculty of Technology & Engineering, The Maharaja Sayajirao University of Baroda, Vadodara, Gujarat

Priya, V. Siridi Sai College of Engineering, Bangalore
E-mail: ludayasimha@yahoo.com; vaishudurga@yahoo.com

Puttaswamy. Department of Chemical Engineering, BMS College of Engineering, Bangalore
E-mail: moodala@gmail.com

Rai, Archita. UP Rajarshi Tandon Open University, Allahabad, U.P.
E-mail: archita_rai@yahoo.com

Rajput, Swati Singh. College of Technology, G.B. Pant University of Agriculture & Technology, Pantnagar – 263 145, U.S. Nagar, Uttarakhand

Ramanathan, A.L. School of Environmental Sciences, Jawaharlal Nehru University, New Delhi

Rastogi, Akanksha. Deptt. of Environmental Sciences, College of Basic Science, G.B. Pant University of Agriculture & Technology, Pantnagar – 263 145, U.S. Nagar, Uttarakhand
E-mail: rastogi34002@gmail.com

Samal, L. Animal Genetics Division, Indian Veterinary Research Institute, Izatnagar – 243 122

Shankhwar, Anil K. Department of Environmental Science, College of Basic Sciences and Humanities, G.B. Pant University of Agriculture & Technology, Pantnagar – 263 145, U.S. Nagar, Uttarakhand

Sharma, P. School of Environmental Sciences, Jawaharlal Nehru University, New Delhi

Sharma, R.C. Department of Environmental Sciences, H.N.B. Garhwal University Srinagar-Garhwal, Uttarakhand
E-mail: mishra.nih@gmail.com

Sharma, Sanchita. MBA from Manipal University
E-mail: sanchita2305@yahoo.co.in

Shrimali, N.J. Associate Professor, The M.S. University of Baroda, Vadodara, Gujarat

Simha. L.U. Department of Civil Engineering, BMS College of Engineering, Bangalore

Singh, V.B. School of Environmental Sciences, Jawaharlal Nehru University, New Delhi
E-mail: virendrasinghjnu@gmail.com

Singh, Vir. Department of Environmental Science, College of Basic Sciences and Humanities, G.B. Pant University of Agriculture & Technology, Pantnagar – 263 145, U.S. Nagar, Uttarakhand
E-mail: drvirsingh@rediffmail.com, sinvir@gmail.com

Srivastava, R.K. Departments of Environmental Sciences, College of Basic Science, G.B. Pant University of Agriculture & Technology, Pantnagar – 263 145, U.S. Nagar, Uttarakhand

Sujatha, V. University of Petroleum and Energy Studies, Dehradun, Uttarakhand

Swamy, Raju Narayan, IAS. Secretary to Government of Kerala
E-mail: narayan5@ias.nic.in,
raju.narayanswamy40@gmail.com

Tyagi, Shweta. College of Biotechnology, Sardar Vallabhbhai Patel University of Agriculture and Technology, Meerut

Part-I

CLIMATE CHANGE PERSPECTIVES AND ISSUES

Chapter 1

Environmental Issues in India: A Perspective

A.K. Dikshit and Raj Kumar

Introduction

Environmental pollution is one of the most serious problems facing humanity and other life forms on our planet earth. The environmental problems in India are also growing rapidly due to economic development and fast increase in population. India's present population of 1.2 billion people is putting a strain on the environment, infrastructure and the country's natural resources. Industrial pollution, soil erosion, deforestation, rapid industrialization, urbanization and land degradation are some of the major environmental problems that need to be properly addressed.

Air Pollution

There are four reasons for air pollution, namely emissions from vehicles, thermal power plants, industries and refineries. Coal meets more than half of the country's energy need and is used for nearly three-quarters of electricity generation. However, Indian coal has high ash content and burning this resource causes environmental degradation. Reliance on coal as major source has led to a nine-fold jump in carbon emissions over the past forty years. Vehicle emissions are responsible for 70 per cent of the country's air pollution. Air pollution from vehicle exhaust and industry is a worsening problem for India. Exhaust from vehicles has increased eight-fold over the levels twenty year ago while industrial pollution has risen four times over the same period. Air quality is worst in big cities like Chennai, Delhi and Kolkata etc. Exhaust from vehicles, dust from construction debris, industrial waste, burning of municipal and garden waste are all on the rise. The levels of respirable suspended particulate matter (RSPM) or dust in the air have continued to increase in the recent past. The toxic value of PM 2 is such that metals like lead present in it get inhaled deeper into lungs and deposit there. New Delhi adds 1,000 new cars a day to the existing four million registered in the city. Pollution levels of nitrogen oxides have been increasing in the city to dangerous levels, which is a clear sign of pollution from vehicles. With pollution level rising across the country, India revised the national ambient air quality standards in November 2009. The ambient air quality standards in India, which is equivalent to the European level, exceeds the standard prevalent in the US.

River Water Pollution

A growing number of bodies of water in India are unfit for human use. It is estimated that around three billion liters of waste are pumped into Delhi's river Yamuna each day. Only 55 percent of the 15 million Delhi residents are connected to the city's sewage system. The remainder flushes their bath water, wash water, and everything else in the pipes and into drains, most of them empty into the Yamuna. Much of the river pollution problems in India are from untreated sewage. Samples taken recently from the Ganges River near Varanasi show that levels of fecal coliform, a dangerous bacterium that comes from untreated sewage, were 3,000 percent higher than what is considered safe for bathing.

Municipal Solid Waste (MSW)

India's urban population is slated to increase from the current 330 million to about 600 million by 2030. The challenge of managing municipal solid waste (MSW) in an environmentally and economically sustainable manner is bound to assume gigantic proportions. The country has over 5,000 cities and towns, which generate about 40 million tones of MSW per year. Until now, MSW management has been considered to be the sole responsibility of urban governments, without the participation of citizens and other stakeholders.

Mining Pollution

A report prepared by the Centre for Science and Environment (CSE) has made extensive analysis of environment degradation and pollution due to mining. The report states that in 2005-2006 alone 1.6 billion tones of waste and overburden from coal, iron ore, limestone and bauxite, were adding to environment pollution. With the annual growth of mining at 10.7 percent and many awaiting approval of the Centre, the pollution would increase manifold in the coming years.

Greenhouse Gas Emissions

India is fifth most carbon emitting country behind US, China, Russia and Japan. Carbon emissions have grown nine-fold over the past forty years. Carbon emissions are forecast to grow 3.2 per cent per annum till 2020. To put this in perspective, carbon emissions levels are estimated to increase by 3.9 per cent for China and by 1 per cent for the United States. In India, carbon dioxide (CO_2) emissions are increasing due to use of fossil fuel. According to the UN Framework Convention on Climate Change, the most industrialized countries are mainly responsible for causing climate change.

Aerosol Pollution

Latest findings by NASA and Stanford University indicate that aerosol pollution will slow down winds, impacting normal rainfall pattern in tropical countries. The unique combination of meteorology landscape (relatively flat plains framed by Himalayas to the north and an open ocean to the south), and the large population maximize the effects of aerosol pollution in India.

Nuclear Energy

India currently gets only a fraction of its electricity from nuclear power. India can fight global warming with clean energy. Despite the advantage of nuclear energy as a clean energy, the big concern is the waste resulting from nuclear reaction, causing pollution of radioactivity. Nuclear waste is also a problem in that the spent nuclear fuel has no safe place to be stored right now.

Agrochemical Pollution

Much of the increase in global agricultural production over the last few decades has come about through the adoption of high-input farming systems. Needless to say that every one has realized the extent to which pests harm crops, cause damage and transmit diseases to both humans and domestic animals. The use of pesticides to exterminate the harmful pests is not a new concept and is still by and far the best method because of ease and rapidity with which insect pests are controlled. But due to non-prescriptive and over use or misuse of these chemical pesticides it may result in harmful effect on the environment and their residues may widely distribute. The very purpose of good agricultural practices is defeated. Though the chemical free agriculture is gaining support but it is still not able to respond to the need for producing massive amount of food. Therefore, pesticides are to stay for long to achieve food security and food safety. A framework policy is required through education, awareness and training for a holistic food chain approach that would recognize that responsibility for supplying safe and nutritious food lies with all the stakeholders in food production and protection and residue free commodities from farm to fork. This further requires adherence of risk characterization and assessment principles. Besides, pesticide regulation is another important aspect in minimizing hazards and promoting the judicious use. Regulation activities are shared by different agencies and governed by different acts, regulations, guidelines, directives and bylaws. Although it is a complex process, regulations at all levels work together towards the common goal, helping to protect citizens from any risk posed by pesticides and ensuring that pest control products do what they claim to do on the label. The food residue exposure assessment is conducted to set the MRLs for pesticides on food. Dietary risk assessments are also carried out to assess the potential daily intake of pesticide residues from all possible food sources. However, dietary risk assessments take into account the different eating patterns. Recently, Government of India has established a single window Food Safety Standards Authority of India under the enactment of Food Safety and Standards Act 2006 to look into the food safety and risk assessment processes and procedures to improve the quality of life. Further, safe and wise use of policy for pesticides, scientific standards like ADI, TMDI and MRLs, risk assessment methodology, surveillance and enforcement for food and body burden, and review and re-reviews of various processes have become essential.

Impact of Use and Abuse of Agrochemicals

The presence of pesticides in the environment, *viz.* soil, air, water, plant, microorganisms, animal and human tissues etc. has been demonstrated the world over. Despite the fact that consumption of pesticides in India is very low as compared to Japan, Korea, USA etc., yet a wide spread contamination of the different environmental components is reported. The persistent misuse and abuse of pesticides reflects poorly on their occupation and safety standards, regulation and enforcement, and knowledge of the hazards. This has resulted in situations posing health risks to humans and the other forms of life and the environmental components. Today's focus is to generate increased farm incomes. The developing countries use about 20 per cent of the globally consumed pesticides. The higher proportion of human poisonings and deaths owing to pesticidal use reflects the existing inadequacies in safety standards. Some of the specific factors contributing to acute poisoning include the use of pesticides in concentrations in excess of requirements, poor quality of spray equipments and their poor maintenance and use of unknown pesticidal cocktails. The credits of the pesticides include economic potential in terms of increased production of food and fiber and amelioration of vector born diseases.

A Concern for Contamination

Various transport and conversion processes take place after the application of agrochemicals or other chemicals. Thus, these enter various environmental components via soil percolation, surface run-off, air drift etc. and affect the non-target species, producing diverse effects on the ecosystem and initiating trophic interactions. Many of these eventually end up in ground water and their residues or transformation products may remain for long periods. Only a few microorganisms are able to degrade these. Rivers and streams also receive the toxic wastes generated on land. The various safety concerns are associated with: (i) active ingredient, (ii) contaminants that exist as impurities along with the active ingredient, (iii) additives that are mixed with the active ingredient, and (iv) degradates formed during physical, chemical or biological transformation of the active ingredient.

Steps to Reduce Pollution

The Centre has in September, 2009 announced "green" bonus and dividends for states maintaining their forest cover. A recent report "Forest and Tree Cover" prepared by the Ministry of Environment and Forests says that from 1995 to 2005, the carbon stocks stored in the forests and trees have increased from 6,245 million tones to 6,662 million ones, registering an annual increment of 38 million tones of carbon or 138 met. of carbon dioxide. India can tackle the issue of global warming effectively if taxes on employees and employers are reduced and put a tax on pollution. The more carbon dioxide one emits the more he pays in taxes". Further, following the rules, regulations and norms as prescribed, and using the active ingredients and formulations which are least toxic and most effective with less residence time and highly dissipative will reduce environmental load.

References

Anonymous. 2001. Pesticide Residues in Food – 2000. *FAO Plant Protection and Production Paper*, FAO.

Anonymous. 2002. Agrochemical Usage. The process of Risk Assessment, Evaluation and Management Ensuring Consumer Safety, Rallis Agrochemical Information Service.

Anonymous. 2004. Hazards associated with pesticides. *Outlooks on Pest Management*, 1(3):105-108.

Anonymous. 2010. WITTS information on Science and Technology Applications (WISTA). *Environment Accounting*. 29:2-19.

Bates, J.A.R. 1991. Perspective on Pesticide Residues in Food. In: *Pesticide Chemistry – Advances in International Research Development and Legislation*. Ed. Helmut Frelize. ISBN 3-527-28111, Weinheim.

Dikshit, A.K. and Jindal, T. 2004a. Exposure assessments across food chemicals. Paper presented at International Conference on Soil and Ground Water Contamination, Risk Assessment And Remedial Measures, pp. 22.

Dikshit, A.K. and Jindal, T. 2004b. Risk assessment of pesticides in soil water- continuum, Paper presented at International Conference on Soil and Ground Water Contamination, Risk Assessment and Remedial Measures, 33.

Dikshit, A.K., Pachauri, D.C. and Jindal, T. 2003. Maximum residue limit and risk assessment of beta-cyfluthrin and imidacloprid on tomato (*Lycopersicon esculentum* Mill). *Bull. Environ. Contam. Toxicol.,* 70:1143-115.

Handbook of Pesticides, Methods of Pesticide Residue Analysis (Eds: LML Nollet and H.S. Rathore); CRC Press, Tylor and Francis Group, Boca Raton.

Rai, Mangala. 2004. Inaugural Address, National conference on Current Issues of Plant Sciences: Health and Environmental Aspects, October 8, 2004, Agra.

Renwick, G.A. 2002. Pesticide residue analysis and its relationship to hazard characterization (ADI/ARfD) and intake estimations (NEDI/NESTI). *Pest. Manag. Sci,.* 1073 – 1082.

WHO. 1997. *Guidelines for Dietary Intake of Pesticide Residues.* Programme of Food Safety and Food Aid, FAO, Rome.

Chapter 2

Global Warming and Climate Change: A Perspective

Chinmoy Mishra and Lipismita Samal

Introduction

Climate changes pose formidable challenge to the development of livestock sector in India. Global warming is the increase in average temperature of the earth's near-surface air and ocean since the mid-20th century and its projected continuation. Global surface temperature increased 0.74 ± 0.18 °C (1.33±0.32°F) during the 100 years ending in 2005. The Intergovernmental Panel on Climate Change (IPCC) concludes that most of the temperature increases since the mid-twentieth century is 'very likely' due to the increase in anthropogenic greenhouse gas concentrations. Natural phenomena such as solar variation and volcanoes probably had a small warming effect from pre-industrial times to 1950 and a small cooling effect from 1950 onward. Increasing global temperature will cause sea levels to rise and will change the amount and pattern of precipitation, likely including an expanse of the subtropical desert regions. Other likely effects include increases in the intensity of extreme weather events, changes in agricultural yields, modifications of trade routes, glacier retreat, species extinctions and increases in the ranges of disease vectors.

Greenhouse Gas Emissions

Gases that trap heat in the atmosphere are often called 'greenhouse gases'. Some greenhouse gases such as carbon dioxide occur naturally and are emitted to the atmosphere through natural processes and human activities. Other greenhouse gases (*e.g.*, fluorinated gases) are created and emitted solely through human activities. The principal greenhouse gases that enter the atmosphere because of human activities are

- *Carbon Dioxide (CO_2)*: Carbon dioxide enters the atmosphere through the burning of fossil fuels (oil, natural gas, and coal), solid waste, trees and wood products, and also as a result of other chemical reactions (*e.g.*, manufacture of cement). Carbon dioxide is also removed from the atmosphere (or 'sequestered') when it is absorbed by plants as part of the biological carbon cycle.
- *Methane (CH_4)*: Methane is emitted during the production and transport of coal, natural gas, and oil. Methane emissions also result from livestock and other agricultural practices and by the decay of organic waste in municipal solid waste landfills.
- *Nitrous Oxide (N_2O)*: Nitrous oxide is emitted during agricultural and industrial activities, as well as during combustion of fossil fuels and solid waste.
- *Fluorinated Gases*: Hydrofluorocarbons, perfluorocarbons, and sulfur hexafluoride are synthetic, powerful greenhouse gases that are emitted from a variety of industrial processes. Fluorinated gases are sometimes used as substitutes for ozone-depleting substances (*i.e.*, Chlorofluorocarbons, Hydrofluorocarbons, and halothanes). These gases are typically emitted in smaller quantities, but because they are potent greenhouse gases, they are sometimes referred to as High Global Warming Potential gases ('High GWP gases').

Greenhouse Effect

The causes of the recent warming are an active field of research. The scientific consensus is that the increase in atmospheric greenhouse gases due to human activity caused most of the warming observed since the start of the industrial era, and the observed warming cannot be satisfactorily explained by natural causes alone. The greenhouse effect was discovered by Joseph Fourier in 1824 and first investigated quantitatively by Svante Arrhenius in 1896. It is the process by which absorption and emission of infrared radiation by atmospheric gases warm a planet's lower atmosphere and surface (Figure 2.1). Existence of the greenhouse effect as such is not disputed. Naturally occurring greenhouse gases have a mean warming effect of about 33°C (59°F), without which earth would be uninhabitable. On earth the major greenhouse gases are water vapor, which causes about 36-70 per cent of the greenhouse effect (not including clouds); carbon dioxide (CO_2), which causes 9-26 per cent ; methane (CH_4), which causes 4-9 per cent ; and ozone, which causes 3-7 per cent.

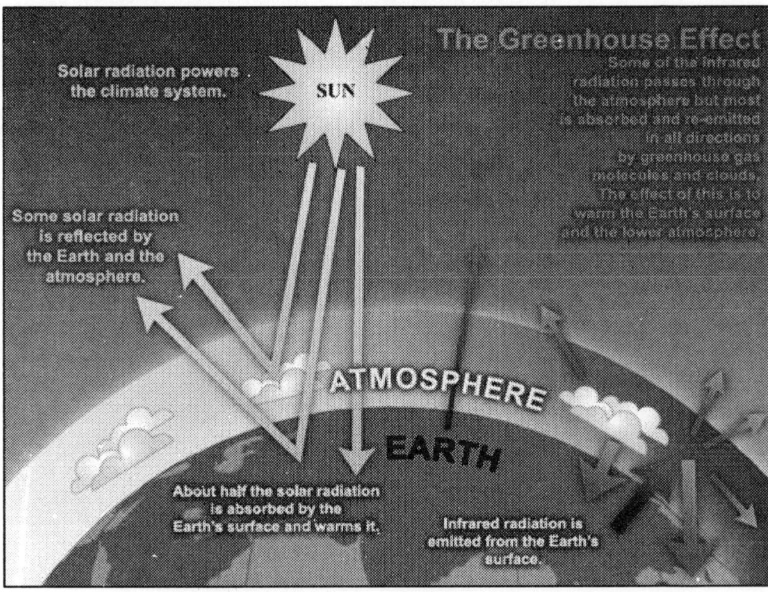

Figure 2.1: The Greenhouse Effect

Human activity since the industrial revolution has increased the atmospheric concentration of various greenhouse gases, leading to increased radioactive forcing from CO_2, methane, tropospheric ozone, CFCs and nitrous oxide. The atmospheric concentrations of CO_2 and methane have increased by 36 per cent and 148 per cent respectively since the beginning of the industrial revolution in the mid-1700s. Fossil fuel burning has produced approximately three-quarters of the increase in CO_2 from human activity over the past 20 years. Most of the rest is due to land-use change, in particular deforestation. CO_2 concentrations are expected to continue to rise due to ongoing burning of fossil fuels and land-use change. The rate of rise will depend on uncertain economic, sociological, technological, and natural developments. The IPCC Special Report on Emissions Scenarios gives a wide range of future CO_2 scenarios, ranging from 541 to 970 ppm by the year 2100 (Figure 2.2). Fossil fuel reserves are sufficient to reach this level and continue emissions past 2100 if coal, tar sands or methane clatharates are extensively exploited.

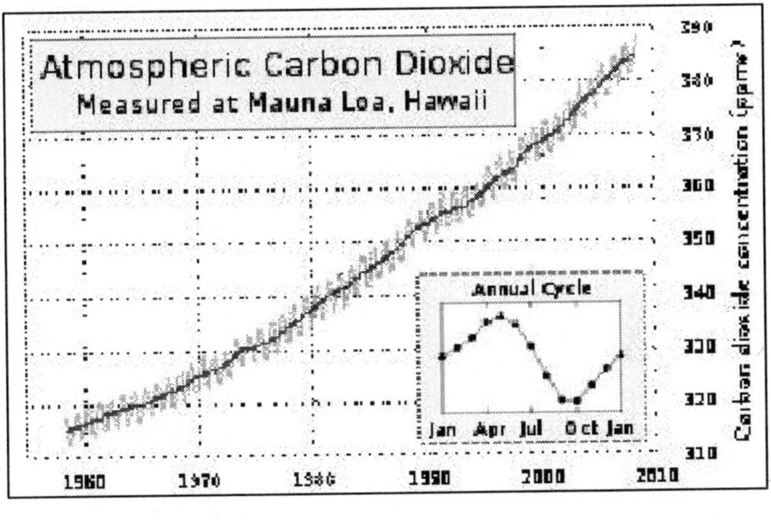

Figure 2.2: Measurement of Atmospheric Carbon Dioxide

Effects of Climate Change on Human Health

Throughout the world, the prevalence of some diseases and other threats to human health depend largely on local climate. Extreme temperatures can lead directly to loss of life, while climate-related disturbances in ecological systems, such as changes in the range of infective parasites, can indirectly impact the incidence of serious infectious diseases. In addition, warm temperatures can increase air and water pollution, which in turn harm human health. Human health is strongly affected by social, political, economic, environmental and technological factors, including urbanization, affluence, scientific developments, individual behavior and individual vulnerability (*e.g.*, genetic makeup, nutritional status, emotional well-being, age, gender and economic status). The extent and nature of climate change impacts on human health vary by region, by relative vulnerability of population groups, by the extent and duration of exposure to climate change itself and by society's ability to adapt to or cope with the change.

Climate change may increase the risk of some infectious diseases, particularly those diseases that appear in warm areas and are spread by mosquitoes and other insects. These 'vector-borne' diseases include malaria, dengue fever, yellow fever, and encephalitis. Also, algal blooms could occur more frequently as temperatures warm particularly in areas with polluted waters in which case diseases (such as cholera) that tend to accompany algal blooms could become more frequent. Higher temperatures, in combination with favorable rainfall patterns, could prolong disease transmission seasons in some locations where certain diseases already exist. In other locations, climate change will decrease transmission via reductions in rainfall or temperatures that are too high for transmission. Though average global temperatures are expected to continue to rise, the potential for an increase in the spread of diseases will depend not only on climatic but also on non-climatic factors, primarily the effectiveness of the public health system (WHO, 2003). The IPCC has noted that the global population at risk from vector-borne malaria will increase by between 220 million and 400 million in the next century (IPCC, 2007). However, socioeconomic factors such as public health measures will play a large role in determining the existence or extent of such infections. Water-borne diseases may increase where warmer air and water temperatures combine with heavy runoff from agricultural and urban surfaces, but may be largely contained by standard water-treatment practices.

Effects of Climate Change on Air Quality

Climate change is expected to contribute to some air quality problems (IPCC, 2007). Respiratory disorders may be exacerbated by warming-induced increases in the frequency of smog (ground-level ozone) events and particulate air pollution. Ground-level ozone can damage lung tissue, and is especially harmful for those with asthma and other chronic lung diseases. Sunlight and high temperatures, combined with other pollutants such as nitrogen oxides and volatile organic compounds, can cause ground-level ozone to increase. Another pollutant of concern is particulate matter also known as particle pollution or PM. When breathed in, these particles can reach the deepest regions of the lungs. Particle pollution is the main cause of visibility impairment (haze) in the nation's cities and national parks. Climate change may indirectly affect the concentration of PM.

Effects of Climate Change on Livestock Health

The major impacts of climate change on livestock health will be via vector-borne diseases. Increasing temperatures have supported the expansions of vector populations into cooler areas, either into higher altitude system (eg. Malaria and livestock tick-borne diseases) or into more temperate zones (eg. outbreak of bluetongue diseases in Northern Europe). Changes in rainfall pattern can also influence an expansion of vectors during wetter years. This may lead to large outbreak of disease, such as those seen in East Africa due to Rift Valley Fever virus, which is transmitted by a wide variety of biting insects. Tse-tse transmits African trypanosomes widely in livestock (ruminants, horses and pigs) which are very sensitive to environmental change. Predictions of climate and population change on tse-tse density indicates that tse-tse population and animal trypanosomiasis will decrease most in semi-arid and sub-humid zones.

Beyond vector-borne diseases, helminth infections, particularly of small ruminants will be greatly influenced by changes in temperature and humidity. Climate changes could also influence disease distribution indirectly through changes in the distribution of livestock. Areas becoming more arid would only be suitable for camels and small ruminants. If these species are forced to aggregate around water points, the incidence of parasitic diseases could increase. The most important adaptive trait of tropical livestock is disease resitance. Particularly for trypanotolerant breeds, climate change my decrease the importance of this trait in sub humid zones. Climatic changes, mediated through changes in crop and livestock practices, could also influence the distribution and impact of several diseases such as malaria across systems and schistosomiasis and lymphatic filariasis in irrigated systems (Patz and Confalonieri, 2005). Climate change is bound to have further impacts on heat-related mortality and morbidity and on the incidence of climate-sensitive infectious diseases (Patz *et al.*, 2005), and these may be considerable.

Other Environmental Effects

Although it is difficult to connect specific weather events to global warming, an increase in global temperatures may in turn cause broader changes, including glacial retreat, arctic shrinkage, and worldwide sea level rise. Changes in the amount and pattern of precipitation may result in flooding and drought. There may also be changes in the frequency and intensity of extreme weather events. Other effects may include changes in agricultural yields, addition of new trade routes, reduced summer stream-flows, species extinctions, and increases in the range of disease vectors. Some effects on both the natural environment and human life are, at least in part, already being attributed to global warming. A 2001 report by the IPCC suggests that glacier retreat, ice shelf disruption, sea level rise, changes in rainfall patterns, and increased intensity and frequency of extreme weather events are attributable in part to global warming (Figure 2.3).

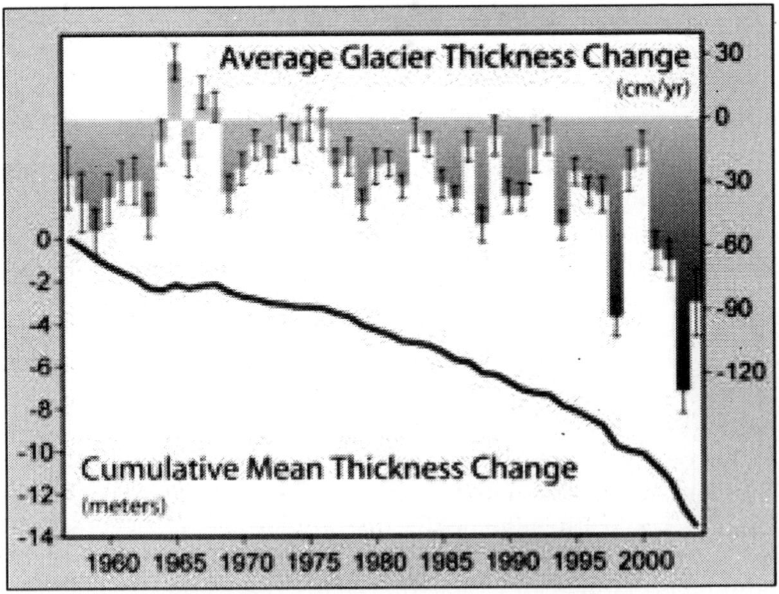

Figure 2.3: Average Glacier Thickness

Other expected effects include water scarcity in some regions and increased precipitation in others, changes in mountain snow-pack, and adverse health effects from warmer temperatures. Social and economic effects of global warming may be exacerbated by growing population densities in affected areas. Temperate regions are projected to experience some benefits, such as fewer deaths due to cold exposure. Additional anticipated effects include sea level rise of 180 to 590 millimeters (0.59 to 1.9 ft) in 2090-2100 relative to 1980-1999, repercussions to agriculture, possible slowing of the thermohaline circulation, reductions in the ozone layer, increasingly intense (but less frequent) hurricanes and extreme weather events, lowering of ocean pH, and the spread of diseases such as malaria and dengue fever.

Conclusion

Global warming is a uniform global issue when it comes to the contribution of individual countries to global warming and the impact that global warming has and will have on individual countries. Similarly, all 6.6 billion people on earth make different contributions to global warming and are differently vulnerable to its impacts. Within each country wealthier individuals usually contribute more to global warming and have more capacity to withstand its negative impacts. In other words, poor countries and poor people within each country bear a lesser responsibility for the changes that are and will occur and are much more vulnerable to the negative impacts of global warming. On Wednesday 16th February 2005, the Kyoto Protocol finally came into force. It aims to tackle global warming by setting target levels for nations to reduce greenhouse gas emissions worldwide that will help to save our earth and its habitants.

References

IPCC Report. 2007. *Historical Overview of Climate Change Science*. IPCC WG1 AR4 Report, Chapter1, p. 97.

IPCC. 2001. Intergovernmental Panel on Climate Change. Climate change (c.f. The Times of India, February 21, 2001, New Delhi, Editorial, p. 12).

IPCC. 2007. Climate Change 2007: Impacts, adaptation, and vulnerability. Contribution of Working Group II to the Third Assessment Report of the Intergovernmental Panel on Climate Change [Parry, Martin L., Canziani, Osvaldo F., Palutikof, Jean P., van der Linden, Paul J., and Hanson, Clair E. (eds.)].

Cambridge University Press, Cambridge, United Kingdom, p. 1000.

Patz, J.A. and Confalonieri, U.E.C. 2005. Human health: ecosystem regulation of infectious diseases. Chapter 14 in Ecosystems and Human well-being: volume 1, current State and Trends. The Millenium Ecosystem Assessment.

Patz, J.A., Campbell-Lendrum, D., Holloway, T. and Foley, J.A. 2005. Impact of regional climate change on human health. *Nature,* 438: 310-317.

WHO. 2003. World Health Organization. Climate change and human health - risks and responses. McMichael, A.J., Campbell-Lendrum, D.H., Corvalán, C.F., Ebi, K.L., Githeko, A., Scheraga, J.D. and Woodward, A. p. 322.

Chapter 3

Global Warming: An Alarming Issue

Aali Pant and Swati Singh Rajput

The increase in the earth's average temperature can further cause other alterations, including an increasing sea level and alterations in the quantity and model of precipitation. These alterations may raise the incidence and concentration of tremendous weather proceedings, such as floods, famines, heat waves, storms, and cyclones. Global warming is caused by green house gases, which trap i the sun's infrared rays in the earth's atmosphere, which n turn heat up the earth's atmosphere. The global warming has happened in the past few years and is evident from the rise in mean temperature of the earth's atmosphere and the melting of the polar ice caps. The effect of global warming can be felt on seasons too with the summers getting longer than the winters. The global warming is also responsible for the introduction of some new diseases. Global warming is also effecting the crop production, as the crops are getting destroyed by the sudden change in temperatures or sudden onset of rains. Thus global warming has affected the life socially; culturally as well as politically.There are few benefits of global warming as well. For instance, with increase in global temperature the Arctic Ocean would be completely open and hence a shorter trade route from Atlantic to Asia will be available. A study shows that global warming will also reduce the frequency of occurrence of hurricanes. Though there are a few benefits of global warming, but the negative effects of it cannot be neglected. To understand the effects of increasing temperature practically, investigations were carried out on a pond soil of Pantnagar. It was found that regardless of high clay content and low permeability of about.0000266 cm/s the water of the pond was completely lost due to evaporation in time duration of nearly 25 days in the month of March itself. This effect has never been seen before.On studying the crop season, *i.e.* the sowing and reaping time of the various crops being cultivated in Pantnagar, it was observed that there is a shift in the sowing and reaping period of various crops like wheat etc. The effects of global warming are very large in number and still there are so many that are still to be found out. But recently the problem has become visible and evident due to happening of the events that were before just talked off.

Introduction

Rapid industrialization and fast population growth have led us to an unwanted danger of environment problem. Science has given us atomic energy, comfortable communication systems etc. for the betterment of the society but at the same time the life has become uncomfortable and tuneful due to the presence of same devices. The pollution is accompanied by the gradual rise in global atmospheric temperature, which is known as global warming. Carbon dioxide is the key factor in causing Global warming. The indiscriminate burning of the fossil fuels like coal, oil and natural gases, releases huge quantities of gases into atmosphere. The industrialized nations have emitted gaseous pollutants more than the developing nations. The fact that the current decade is the hottest period of the century is an indication that the globe is warming up. This will lead to significant adverse effects on the world's natural system and human population over the next century.

Dynamism of the Earth's Temperature

Climate of the globe has always changed; it is the rate of change that is of current concern. The climate of planet Earth is unstable. Our evolutionary origins lie in the warm, relatively benign climate of equatorial Africa, but our ancestors battled the cold, harsh, and unforgiving climate of the last ice age in order to spread across the planet. Some 10,000 years ago, however, the ice age ended.

We developed agriculture, civilization, industry, and technology generally in a global climate that was warm, pleasant, and mostly predictable. Regional climates have changed, sometimes drastically and disastrously for local human populations, but by and large the global climate has not dealt any significant long-term blows to the spread and development of human civilization. One of the most significant accomplishments of our species is the discovery of fossil fuels and the means of turning the energy trapped within them into heat, transportation, and the basis for manufacturing and construction.

This discovery and the global industrial revolution that followed changed the world forever for our species. In general, fossil fuels are a legacy bequeathed to us by the biosphere of the distant past. Since the 1800s, we've been burning vast quantities of these fossil fuels to power our developing technological and global civilization. As a result, we've been releasing the CO_2 trapped in the fuels in the form of energy-rich organic molecules back into the atmosphere, increasing the atmospheric concentration of CO_2. The released CO_2 has resulted in Greenhouse Effect and thus leads to an increase in temperature at a very high rate. Since the 1800s, CO_2 concentrations worldwide have increased from approximately 280 ppm (or 0.028 per cent) to around 365 ppm (0.0365 per cent).

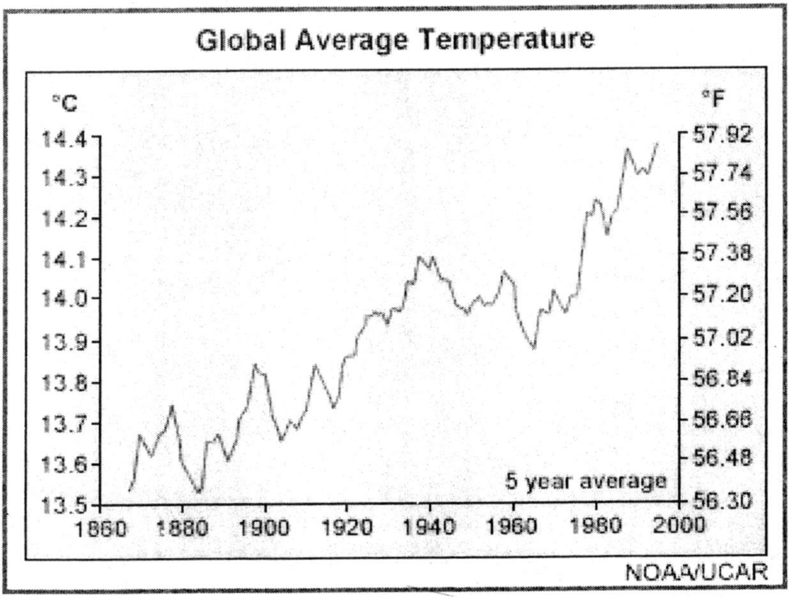

Figure 3.1

The increase seems trivial, but it also means that some 3 gigatons (3 billion metric tons) of CO_2 are being added to the atmosphere every year. As a result, the Earth's average temperature rose about 0.6° Celsius (1.1° Fahrenheit) in the 20th century.

Global Warming vs Climate Change

Global warming and climate change are not synonymous phrases. The two are not alike. Global warming is the increase of the Earth's average surface temperature due to a build-up of greenhouse gases in the atmosphere. While climate change is a broader term that refers to long-term changes in climate, including average temperature and precipitation.

Global Warming Causes

The earth's climate is dynamic and always changing through a natural cycle. What the world is more worried about is that the changes that are occurring today have been speeded up because of industrialization. The causes of climate change can be divided into two categories– those that are due to natural causes and those that are created by man.

Natural Causes

Natural causes are causes created by nature. One natural cause is a release of methane gas from arctic tundra and wetlands. Methane is a greenhouse gas. A greenhouse gas is a gas that traps heat in the earth's atmosphere. Another natural cause is that the earth goes through a cycle of climate change. This climate change usually lasts about 40,000 years.

Man-made Causes

Man-made causes probably do the most damage. When fossil fuels are burned they give off a green house gas called CO_2. The greenhouse gases just like a greenhouse used for growing plants, admit the high wavelength infrared waves of the sun. But when these waves are reflected back their wavelength reduces. The greenhouse gases acting like a glass sheet, do not allow these waves to escape, thus resulting in an increase in temperature. The major source of carbon dioxide is the power plants. These power plants emit large amounts of carbon dioxide produced from burning of fossil fuels for the purpose of electricity generation. Coal is the major fuel that is burnt in these power plants. Coal produces around 1.7 times as much carbon dioxide per unit of energy when flamed as does natural gas and 1.25 times as much as oil. The second major greenhouse gas after carbon dioxide, which causes global warming, is Methane. Methane is more than *20 times* as effectual as CO_2 at entrapping heat in the atmosphere. Methane is naturally in the ground. When coal or oil is mined you have to dig up the earth a little. When the fossil fuels are dug the methane is released as well. Methane is also obtained from resources such as rice paddies, bovine flatulence, bacteria in bogs and fossil fuel manufacture. When fields are flooded, anaerobic situation builds up and the organic matter in the soil decays, releasing methane to the atmosphere. Another major cause of global warming is deforestation. Deforestation is to be blamed for 25 per cent of all carbon dioxide release entering the atmosphere, by the cutting and burning of about 34 million acres of trees each year.

Effects of Global Warming

Global warming has many worst effects that can be seen more prominently in future, though its dreadful consequences can also be seen in the present. According to the foreign secretary, Margaret Beckett, "Climate change could spawn a new era of conflicts around the world over water and other scarce resources unless more is done to curb greenhouse gas emissions."

Speaking at the Royal United Services Institute in London, Mrs. Beckett quoted evidence that a conflict was brewing in Ghana where Fulani cattle herdsmen are reportedly arming themselves to take on local farmers in a confrontation over water and land as climate change expands the Sahara desert. The foreign secretary said the Middle East - with 5 per cent of the world's population but only 1 per cent of its water - would be particularly badly affected, with Saudi Arabia, Iran and Iraq particularly hard hit by a drop in rainfall.

She said the Nile could lose 80 per cent of its flow into Egypt, a country which would also be threatened by rising sea levels in the Nile delta, its agricultural heartland, where flooding could displace 2 million people, threatening internal stability. This is a mirror of what's going on in the world. Following pictures of Colorado River Arizona in June, 2002 and in Dec, 2003, a great difference in water level can be easily observed.

Figure 3.2: River Arizona in June 2002 (Top) and in December 2003

Other Consequences of Global Warming

Rapid Changes in Global Temperature

Increased greenhouse gases and the greenhouse effect is feared to contribute to an overall warming of the Earth's climate, leading to a global warming (even though some regions may experience cooling, or wetter weather, while the temperature of the planet on average would rise).

Some of the Facts

1. The five warmest years on record, in order (warmest first) are 2005, 1998, 2002, 2003 and 2004.
2. According to the World Meteorological Organization (WMO), the 1990s was the warmest decade;
3. The 1900s was the warmest century during the last 1,000 years.

However, it is the *rapid* pace at which the temperature will rise that will result in many negative impacts to humans and the environment and this is why there is such a world-wide concern.

Extreme Weather Patterns

Most scientists believe that the warming of the climate will lead to more extreme weather patterns such as

1. More hurricanes and drought;
2. Longer spells of dry heat or intense rain (depending on where you are in the world);
3. Scientists have pointed out that Northern Europe could be severely affected with *colder* weather if climate change continues, as the arctic begins to melt and send fresher waters further south. It would effectively cut off the Gulf Stream that brings warmth from the Gulf of Mexico, keeping countries such as Britain warmer than expected;
4. In South Asia, the Himalayan glaciers could retreat causing water scarcity in the long run.

Ecosystem Impacts

1. Rapid global heating according to a US National Academy of Science warning;
2. Dramatic increase in greenhouse gas emissions;
3. Ozone loss aggravated by global warming;
4. Ozone loss likely to aggravate global warming;
5. Warming of the oceans leads to increased green house gasses;

6. Permafrost thawing will aggravate global warming;
7. Oceanic changes observed that may aggravate the situation;
8. A vicious circle whereby each problem will exacerbate other problems which will feedback into each other;
9. Massive extinction of species will aggravate the environmental crisis;
10. Sudden collapse of biological and ecological systems may occur, but will have a very slow recovery;
11. While effective measures can decrease global warming and other problems the World community has repeatedly failed to establish cooperation.

Rising Sea Levels

Water expands when heated, and sea levels are expected to rise due to climate change. Rising sea levels will also result as the polar caps begin to melt. The *World Watch Institute* reports that "The Earth's ice cover is melting in more places and at higher rates than at any time since record keeping began" (March 6, 2000). Rising sea levels will impact many coastlines, and a large mass of humanity lives near the coasts or by major rivers. Analysis by the World Wildlife Fund has found that many cities are unprepared for climate change effects such as rising sea levels. There are many other effect of global warming which are shocking and frightful like: failing agricultural output; increase in world hunger, changing crops, melting glaciers, worsening floods, creeping deserts etc.

Local Effects of Global Warming

In order to study the local effects of global warming, a survey regarding it was carried out. In our experiment we studied the evaporation rate of water of one of the ponds in Pantnagar Campus. It was seen that the pond that was flourishing with water in the month of February had almost no water by the end of March. The water could be lost either by evaporation or seepage. Thus in order to find the real cause of it we conducted falling head permeability test to determine the permeability of the pond's soil.

It was observed that the permeability of the soil was of the order 3.88182 x 10^$^{-5}$. This indicates the presence of clay in the soil thereby proving that the water was lost due to evaporation and not by seepage.

Laboratory Observations

Table 3.1: Falling Head Permeability Test

Sl.No.	Tube Area a (cm^2)	Container Area A (cm^2)	Length l (cm)	Time t (sec)	Initial Head (h1) cm	Final Head (h2) cm	Coefficient of Permeability, k(cm/s)
1.	3.14	78.539	35	300	85.7	84.8	4.92500x10^-5
2.	3.14	78.539	35	600	84.8	81.9	0.12646x10^-5
3.	3.14	78.539	35	5040	80.9	63.8	6.59400x10^-5
						k_{avg} =	3.88182x10^-5

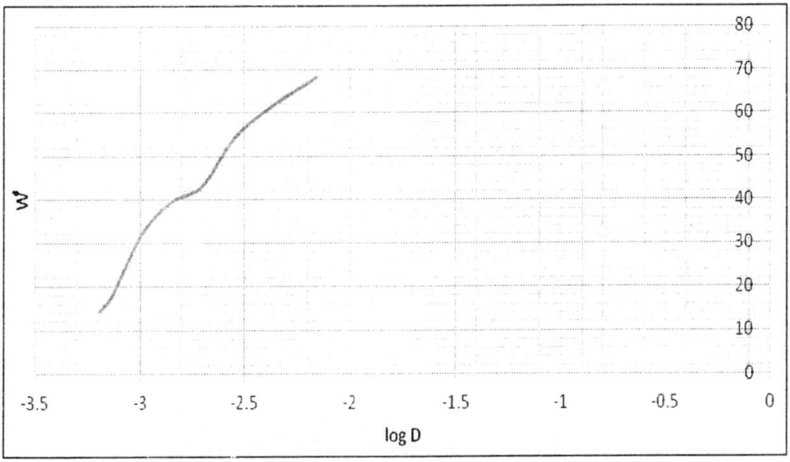

Figure 3.3: Graph Between Percentage Finer and Particle Size

This is a graph plotted between percentage finer and logarithem of particle size. This graph shows that particle size varies between $10^{\wedge -2.15}$ to $10^{\wedge -3.3}$ mm and the size of clay varies from 0.002 to 0.003 mm. this experiment proves that enough quantity of clay was present in the soil.

Survey Conducted in Pantnagar

To study the temperature variation in Pantnagar a survey was conducted on the data collected from CRC (Crop Research Centre). This temperature variation is strictly for Pantnagar.

A graph was prepared between temperature and year to show that the temperature variation in the past few years is very inconsistent. As we can observe from the graph that after 90's temperature variation is very erratic. This discrepancy proves that global warming is occurring and having an adverse effect on Pantnagar too.

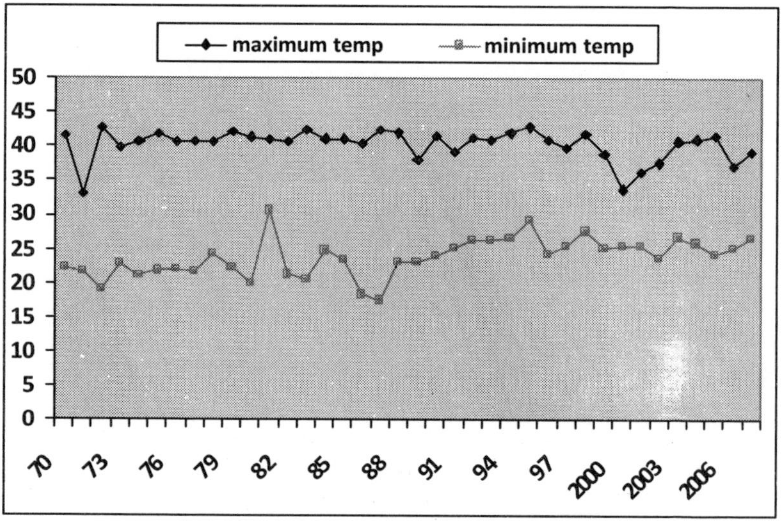

Figure 3.4

Suggestion for Reducing Global Warming

Civil Engineering Aspects

A lot of greenhouse gases and other harmful products are produced during construction projects (for example, in construction of buildings and industries). This can be compensated as well as reduced by adopting sustainable construction methods like bio climatic architecture and green building. The essential requirement in this respect is to adopt the "green building" practice. It is a process to create buildings and infrastructure that minimize the use of resources, reduce harmful effects on the ecology and create healthier environments for people. Green buildings are expected to exhibit a high level of environmental, economic and engineering performance that includes:

1. Energy efficiency and conservation
2. Improved indoor air quality
3. Resource and material efficiency
4. Occupant's health and productivity
5. Proximity to public transport system
6. Improved environmental quality covering air
7. Water, land and ecosystems

Carbon dioxide is an important greenhouse gas. It plays a great role in global warming. There are many factors that are responsible for the production of carbon dioxide. One of the major reasons is cement industry. Following steps should be taken in cement industries to reduce the emission of CO_2:

1. Reduce the consumption of concrete
2. Reduce the cementing materials in concrete mixtures
3. Reduce the clinker factor of cement

Other Measures

1. Replace a regular incandescent light bulb with a compact fluorescent light bulb (cfl)
2. Install a programmable thermostat
3. Move the thermostat down 2° in winter and up 2° in summer
4. Clean or replace filters on the furnace and air conditioner
5. Choose energy efficient appliances when making new purchases
6. Do not leave appliances on standby
7. Defrost old fridges and freezers regularly
8. Replace the old single-glazed windows with double-glazing
9. Get a home energy audit
10. Cover the pots while cooking

11. Use the washing machine or dishwasher only when they are full
12. Take a shower instead of a bath
13. Use less hot water
14. Use a clothesline instead of a dryer whenever possible
15. Insulate and weatherize the home
16. Recycle the organic waste
17. Choose products that come with little packaging and buy refills when it is possible
18. Reuse the shopping bag
19. Reduce waste
20. Switch to green power
21. Buy locally grown and produced foods
22. Buy fresh foods instead of frozen
23. Seek out and support local farmers markets
24. Buy organic foods as much as possible
25. Drive carefully and do not waste fuel
26. Check the tires weekly to make sure they're properly inflated
27. When it is time for a new car, choose a more fuel efficient vehicle
28. Try car sharing
29. Encourage the switch to renewable energy
30. Protect and conserve forest worldwide.

Our Contribution to Society

Global warming is an acute problem that has been affecting the earth for quite a while. Animals have been losing their homes; severe weather has caused lots of damage to homes and communities. Also all the smoking and trash has been polluting the environment and it needs to be stopped. There are some things we can do about it whether we had a position of influence or as an average person.

It is dreadful to see that Mother Earth has been wounded, so it becomes our basic and fundamental duty to protect our environment in the best way we can. We must try to promote the concept of bioclimatic architecture and green building in our country through various means like paper presentation and idea of implementation of bioclimatic architecture on our department's design map and layout. Apart from these, we should try to reduce and recycle the consumption of plastic bags as far as we could.

Conclusion

It is a well known concept that the climate has considerably changed in the past few decades. This is basically due to a sudden wave of industrial revolution in the world. Prior to 18th century, the temperature did not vary the way it varies today. If global warming is not controlled then no doubt humanity will not be lost, but it will not remain as it is today.

References

Chatterjee, A.K. 2009. Sustainable construction and green buildings on the foundation of building ecology, *The Indian Concrete Journal* 5, pp. 27.

Tande, S.N. and Krishnaswamy, K.T. 2009. Greening of concrete industry for environmentally compatible and sustainable structures, *The Indian Concrete Journal* 5. pp. 39.

Chapter 4

Microalgae in the Greenhouse Gas Sequestration

Deeksha Baliyan, Shweta Tyagi and S.K. Bhatnagar

With the imminent threat of anthropogenic climate change in the coming decades, helping to control India's emissions will have to be a global priority. Carbon capture and sequestration (CCS) can play a pivotal role in reducing India's emissions in the future, given India's reliance on coal power and the large coal reserves. This Chapter tries to ascertain the current situation and conditions relevant to carbon capture in India so as to help guide the processes to prepare for large scale adoption if desired in the future. For carbon capture to be undertaken at a significant scale, various pieces will have to fall into place in sync with each other. Biological fixation of carbon dioxide is an attractive option because plants naturally capture and use carbon dioxide as a part of the photosynthetic process. However, because of the relatively small percentage of carbon dioxide in the atmosphere (approximately 0.036 per cent), the use of terrestrial plants is not an economically feasible option. Biosequestration of carbon dioxide using microalgae has emerged as a potential option helping alleviate the trend toward global warming. Microalgae have the advantages of efficient photosynthesis superior to the terrestrial plants, fast proliferation rates, wide tolerance to extreme environments, and potential for intensive cultures. These advantages promise high performance in the reduction of carbon dioxide. Once harvested, microalgae can serve as a product to offset some of the costs that have been incurred. Potential uses for the algae include biodiesel, biofuel for use in electricity production, fodder for livestock, food and chemicals, colorants, perfumes, and vitamins. Ponds are clogged with algae during the summer and this can easily be imagined being literally a 'green' solution that would allow us to continue burning fossil fuels. Unfortunately, getting from here to there is not that easy. To realize workable biological CO_2 fixation systems, selection of optimal microalgae species is vital. The selection of optimal microalgae species depends on specific strategies employed for CO_2 sequestration. In this paper, the criteria used for selecting microalgae species for CO_2 sequestration systems will be discussed, as well as the characteristics of some species which have been tested for use in CO_2 mitigation. The objective of this Chapter is to increase the understanding of the opportunities, issues and challenges regarding carbon sequestration using algae in India regarding the capacity, political structures and policies.

Introduction

Increased CO_2 concentration leading to global warming is receiving a great deal of attention all over the world. Atmospheric increases of carbon dioxide are positively correlated with the amount of fossil fuels being burned. In an effort to retard this increase and, therefore, the greenhouse effect, most industrialized countries have joined in a policy to hold carbon dioxide emissions in the year 2000 to 1990 levels. However, it is not clear if technology exists to achieve this goal. Selecting the most appropriate technology to limit the amount of carbon dioxide entering the atmosphere has been the major focus of research. One option is carbon dioxide capture and subsequent sequestration into the deep ocean, aquifers or depleted oil and gas wells. However, this is an expensive option (potentially more than doubling the cost of electrical generation via fossil fuels) with no opportunity for profit to displace the cost (Stephan *et al.*, 2002).

Biological fixation of carbon dioxide is an attractive option because plants naturally capture and use carbon dioxide as a part of the photosynthetic process. Terrestrial plants sequester vast amounts of carbon dioxide from the atmosphere. However, because of the relatively small percentage of carbon dioxide in the atmosphere (approximately 0.036 per cent), the use of terrestrial plants is not an economically feasible option. On the other hand, discharge gases from heavy industries commonly contain carbon dioxide levels significantly higher than that found in the atmosphere (10 per cent –20 per cent). Therefore, it would be wise to develop strategies to limit this value. Biofixation of carbon dioxide using microalgae has emerged as a potential option. Microalgae have the advantages of efficient photosynthesis superior to C4 plants (those plants that form four carbon stable intermediates in the photosynthetic process; generally associated with agricultural and large terrestrial plants), fast proliferation rates, wide tolerance to extreme environments, and potential for intensive cultures. These advantages promise high performance in the reduction of carbon dioxide. Once harvested, microalgae can serve as a product to offset some of the costs that have been incurred. Potential uses for the algae include biodiesel, biofuel for use in electricity production, fodder for livestock, food and chemicals, colorants, perfumes, and vitamins Stephan *et al.*, 2002). Government of India has drawn a plan for climate change mitigation (NAPCC, 2008) which, among several measures, also suggests enhancement of CO_2 sequestration.

Commercial Values

Some microalgae species, such as *Chlorella, Spirulina* and *Dunaliella* have commercial values. It is expected that commercial profit from biomass production will offset overall operational costs for CO_2 sequestration. *Chlorella* sp. has been studied for use in CO_2 sequestration. For example, Hanagata *et al.* (1992) reported that *Chlorella* sp. can be grown under 20 per cent CO_2 conditions. The species has been used as a health food. CO_2 tolerance of *Dunaliella* sp. also has been examined and the species has been used in the industrial production of β-carotene. Further potential applications of microalgal products are the utilization of secondary metabolite, fertilizer and biofuel production. In addition to CO_2 sequestration, another potential strategy to offset operational cost is to develop multi-functional systems such as waste treatment and aquaculture farms, functions. Since economic feasibility is one of the major issues to realize biological mitigation systems, seeking additional value for the system is an important criterion (Ono and Cuello, 2002).

Cultural Systems

Two distinctive cultural systems have been proposed for CO_2 sequestration with microalgae. One is the open pond system, and the other is the closed photobioreactor system. There is ongoing discussion regarding whether the open pond system or the closed photobioreactor system would be better for CO_2 sequestration. Apparent advantages for utilizing the open pond system are low initial and operational costs. On the other hand, an advantage for the photobioreactor system has a higher potential productivity due to better environmental control and harvesting efficiency. For an open pond system, the size of the area needed to assimilate significant amounts of CO_2 is being criticized. In fact, the typical size of open pond microalgae production systems ranges from 0.2 to 0.4 ha. However, there are already existing large-scale open pond systems. Another strategy, which has never been explored for CO_2 sequestration use, is to build moderately environmentally controlled systems. In such a system, microalgae are grown under a relatively controlled environment, such as a greenhouse and shade. This is an interesting idea since one can control environment inside a greenhouse while construction costs will not be as high as a photobioreactor with a solar collector system. In fact, a similar concept has been applied for wastewater treatment (Ono and Cuello, 2002).

Photosynthesis

Photosynthesis is the formation of carbohydrates in the chlorophyll-containing tissues of plants exposed to light. During photosynthesis in green plants, light energy is captured and used to convert water, carbon dioxide, and minerals into oxygen and energy-rich organic compounds. In many green plants, carbohydrates are the most important direct organic products of photosynthesis. The formation of a simple carbohydrate, glucose, is indicated by the following chemical equation:

$$6\,CO_2 + 12\,H_2O + \text{light} + \text{green plant} = C_6H_{12}O_6 + 6\,O_2 + 6\,H_2O$$

Glucose is then converted in the plant to starch and cellulose (which are polymers of glucose), sucrose, amino acids, proteins, fats, pigments, and other organic compounds. Chemical bonds are broken between the carbon and oxygen (in the CO_2) and between the hydrogen and oxygen (in the water), and new chemical bonds are formed in the organic compounds. More energy is required to break the bonds of CO_2 and H_2O than is released when the organic compounds are formed. This excess bond energy accounts for the light energy stored as chemical energy in the organic compounds formed during photosynthesis. The amount of light, the carbon dioxide concentration, and the temperature are the three most important environmental factors that directly affect the rate of photosynthesis. Plant species and plant physiological state also affect the rate of photosynthesis. The energy efficiency of photosynthesis is defined as the ratio of energy stored as organic compounds to the energy of light absorbed. The maximum efficiency of glucose production is about 31 per cent. Energy stored in the chemical bonds of the organic compounds will be released when the organic compound is combined with oxygen during combustion or when metabolized by an animal or other organism (Stephan *et al.*, 2002).

Effects of Flue Gas on Microalgae

Sulfur oxides, particularly SO_2, can have a significant effect on the growth rates and health of the microalgae. Of greatest concern is the effect SO_2 has on the pH of the microalgal growth medium. When the SO_2 concentration reaches 400 ppm, the pH of the medium can become lower than 4 in less than a day, which significantly affects the productivity of the microalgae. However, if the pH is maintained at 8 using NaOH, the productivity does not decrease. Nannochloris sp. (NANNO0_2) was found to be resistant to 50 ppm SO_2, but without pH control, 300 ppm SO_2 inhibited growth within 20 hours (Stephan *et al.*, 2002).

Nitrogen oxides also comprise a significant portion of power plant flue gas. As with the sulfur oxides, nitrogen oxides can affect the pH of the algal medium, but to a lesser degree. Microalgae have been shown to tolerate and grow in a medium containing 240 ppm NO_x, with pH adjustment. Microalgae are not growth inhibited by the presence of 150 ppm NO. It was also found NANNO02 grew in the presence of 300 ppm NO after a considerable lag time. A point of interest is that the nitrogen oxides can serve as a nitrogen source for the microalgae. NO is absorbed into the medium and oxidized into NO_2 in the presence of oxygen. The greater the oxygen content of the medium, the greater the NO_2 production and microalgal productivity rates. However, the presence of elevated concentrations of oxygen results in algal photorespiration, which inhibits microalgal growth.

When the flue gas is taken as a whole and applied to the microalgae, *i.e.*, not in individual components, the results do not change much. Some microalgal strains have shown tolerance to a medium containing a carbon dioxide concentration of 13.6 per cent, 200 ppm sulfur dioxide, and 150 ppm NO. One marine species (*Tetraselmis* sp.) was cultivated without difficulty while subjected to flue gas containing 14.1 per cent CO_2, 1.3 per cent O_2, 185 ppm SO_x, and 125 ppm NO_x. Other strains of microalgae have shown similar tolerances to carbon dioxide and NO_x, but less tolerance to the sulfur oxides. Another concern with the flue gas is the elevated temperatures. In a commercial application, flue gas from the desulfurization scrubbers would be sent to the CO_2 sequestration ponds for treatment. Temperatures exiting the scrubbers at the Coal Creek Station are approximately 140°F (60°C). Therefore, microalgae would need to exhibit tolerances at temperatures of this magnitude. Although most organisms cannot survive at these higher temperatures, some cyanophycean algae have been shown to grow at 176°F (80°C) (Stephan *et al.*, 2002).

CO_2 Uptake and Algal Growth Rates

To determine the viability of using microalgae as a carbon dioxide sequestration option, carbon dioxide fixation and microalgae growth rates have been studied. It should be noted that current data were taken from microalgae grown in specially constructed photobioreactors (PBRs), or growth chambers, except where it is otherwise specified. The microalgae growth rates that occur in bioreactors and small scale ponds cannot be expected in full-scale operating sequestration ponds. Photosynthetic visible light conversion efficiencies in laboratory experiments are often 20 per cent–24 per cent, whereas maximum conversion efficiencies of higher plants (such as sugar cane) grown at optimum conditions are only 7 per cent–8 per cent. The conversion efficiencies of microalgae are not expected to exceed those of the higher plants. And to achieve those efficiencies would require development of techniques to maintain the original algal strains and ward off invasion by other algae, zooplankton, and infections (Stephan *et al.*, 2002).

Lighting and Nutrient Requirements

The basic nutrient requirements of the microalgae include carbon dioxide, nitrogen and phosphorus along with trace nutrients (minerals and metals). Flue gas supplies the carbon dioxide and has the ability to supply some of the nitrogen, as previously discussed. However, it has been demonstrated that the nitrogen contribution from the flue gas is insufficient to maintain stable growth rates. Therefore, nitrogen, phosphorus, and trace nutrients can be added and maintained as necessary with proper engineering.

The most appropriate light source and method of introduction to the biomass has been the focus of much research. In the case of the PBRs, the design of the PBR is the most important consideration for light penetration and distribution. Tubular PBRs achieve a higher photosynthetic efficiency (PE; defined as how efficiently radiant energy is converted into chemical energy within the microalgae) than flat PBRs. The curved surfaces more evenly distribute the light within the reactor by maximizing the illumination surface-to-volume ratio. Other designs include spiral and helical PBRs (Stephan *et al.*, 2002).

Light condition, especially light intensity, is an important factor because the light energy drives photosynthesis. Typical light intensity requirements of microalgae are relatively low in comparison to higher plants. For example, saturating light intensity of *Chlorella* sp. and *Scenedesmus* sp. is approximately 200 mmol/sec/m^2. Microalgae often exhibit photoinhibition under excess light conditions. Photoinhibition is often suspected as the major cause of reducing algal productivity. The use of a photobioreactor with a solar collector device for the CO_2 mitigation has been explored. Maximum light intensity of 15.7 Wm^{-2} could be attained using the system, and the culture of *Chlorella* sp. could be maintained. The efficiency of light collection and transmission to the algal cells was 8 per cent. In addition, artificial lighting is combined so that lighting is possible when there is no natural sunlight. The use of such novel solar collecting and distributing devices would improve CO_2 sequestration efficiency (Stephan *et al.*, 2002).

High CO_2 Tolerance

Direct utilization of power plant flue gas has been considered for CO_2 sequestration systems. The advantage of utilizing flue gas directly is the reduction of the cost of separating CO_2 gas. Since power plant flue gas contains a higher concentration of CO_2, identifying high CO_2 tolerant species is important. Although CO_2 concentrations vary depending on the flue gas source, 15-20 per cent v/v is typically assumed (Ono and Cuello, 2002).

Several species have been tested under CO_2 concentrations of over 15 per cent. For example, *Chlorococcum littorale* could grow under 60 per cent CO_2 using the stepwise adaptation technique. Another high CO_2 tolerant species is *Euglena gracilis*. Growth of *Euglena gracilis* was enhanced under 5-45 per cent concentration of CO_2. The best growth was observed with 5 per cent CO_2 concentration. However, the species did not grow under greater than 45 per cent CO_2. *Chlorella* sp. UK001 could grow successfully under 10 per cent CO_2 conditions. It is also found that *Chlorella* sp. can be grown under 40 per cent CO_2 conditions. Table 4.1 summarizes the CO_2 tolerance of various species. Note that some species may tolerate even higher carbon dioxide concentrations than listed in the table. Overall, a number of high CO_2 tolerant species have been identified (Ono and Cuello, 2002).

Table 4.1: CO_2 Tolerance of Various Species

Species	*Maximum CO_2 concentration*
Cyanidium caldarium	100 per cent
Scenedesmus sp.	80 per cent
Chlorococcum littorale	60 per cent
Synechococcus elongatus	60 per cent
Euglena gracilis	45 per cent
Chlorella sp.	40 per cent
Eudorina spp.	20 per cent
Dunaliella tertiolecta	15 per cent
Nannochloris sp.	15 per cent
Chlamydomonas sp.	15 per cent
Tetraselmis sp.	14 per cent

Source: Ono and Cuello (2002).

Tolerance of Trace Elements in the Flue Gas

Some researchers considered the effect of trace acid gases on CO_2 sequestration by microalgae, such as NO_x and SO_2. As a source of trace elements, both model flue and actual flue gas have been used. It is found that *Nannochloris* sp. could grow under 100 ppm of nitric oxide (NO). Under 1000 ppm of NO and 15 per cent CO_2 concentration, *Dunaliella tertiolecta* could remove 51 to 96 per cent of nitric oxide depending on the growth condition. *Tetraselmis* sp. could grow with actual flue gas with 185 ppm of SO_x and 125 ppm of NO_x in addition to 14.1 per cent CO_2 (Ono and Cuello, 2002).

High Temperature Tolerance

Since the temperature of waste gas from thermal power stations is around 120°C, the use of thermophilic or high temperature tolerant species is also considered. Themophiles can grow in temperature ranging from 42-100°C. An obvious advantage of the use of thermophilies for CO_2 sequestration is reduced cooling costs. In addition, some thermophiles produce unique secondary metabolites, which may reduce overall costs for CO_2 sequestration. A disadvantage is the increased loss of water due to evaporation. *Cyanidium caldarium*, which can grow under pure CO_2 is a thermophilic species (Ono and Cuello, 2002). Although less tolerant than thermophiles, some mesophiles can still be productive under relatively high temperature. Such species also can be candidate species for the direct use of flue injection.

Marine Microalgae

The use of marine microalgae for biological CO_2 sequestration has been considered. One reason is that seawater could be used directly as a growing media so that maintenance costs of microalgae culture could be reduced. Many CO_2 sources, such as power plants, are located along the coastal area. A number of marine algae species have been tested for CO_2 sequestration applications. Those marine algae species are *Tetraselmis* sp., *Synechococcus* spp., *Chlorococcum littorale*, *Chlamydomonas* sp., *Nannochloropsis salina* and *Phaeodactylum tricornutum* (Ono and Cuello, 2002).

Conclusion

CO_2 sequestration is a critical phenomenon to keep balance of atmospheric carbon dioxide to contribute to climate change mitigation. There are several measures to use in a normal way and to accelerate CO_2 sequestration. The use of microalgae is a unique technology for this purpose. Several species of microalgae offer scope for increasing CO_2 sequestration. Microalgae can assimilate CO_2 within various ranges of concentration from ambient (0.04 per cent) to 100 per cent v/v CO_2 by selecting adequate species. Microalgae-based technology also works under a wide range of thermal conditions, ranging from 25 to 100°C. Adapting microalgae for the use of CO_2 sequestration also has the potential to produce useful byproducts, and could function multi-purposely. In addition, it is an environmentally friendly technology.

References

NAPCC. 2008. National Action Plan on Climate Change. Prime Minister's Council on Climate Change, Government of India. Available at: http://pewclimate.org/international/country-policies/indiaclimate-plan-summary/06-2008.

Ono, E. and Cuello, J.L. 2002. Selection of optimal microalgae species for CO_2 sequestration. http://www.netl.doe.gov/publications/proceedings/03/carbon-seq/PDFs/158.pdf

Stephan, D.J., Shokey, R.E., Moe, T.A. and Dorn, R. 2002. Carbon dioxide sequestering using microalgal systems. http://www.osti.gov/bridge/servlets/purl/882000-St23VC/882000.pdf

Chapter 5

Climate Chaos? A Hysteria and Simple Myth

I.M. Arora

The day before yesterday they called it "global warming". Yesterday they called it "climate change". Today they call it "energy security". Tomorrow they will call it what it is – absolute ridiculous. Climate research institutions know that they must present scary climate forecasts to receive continued funding– no crisis means no funding. The media presents stories of climate disaster to sell their products. Scientific research that suggests climate change is mostly natural and does not receive much of any media coverage. These factors have caused the general public to be seriously misled on climate issues resulting in wasteful expenditures of billions of dollars in an ineffective attempt to control climate. A new study based on computer models has suggested that India is sinking into the Earth's mantle. Scientists have been mystified by the fact that the Himalayan mountain range is still growing, despite the fact that the Indian tectonic plate has been slamming into the Eurasia plate from the past 50 million years. According to a report in *Discovery News*, a new study, led by Fabio Capitanio of Monash University in Australia, based on computer models of the two plates shows that the formation and continued growth of the world's highest mountain range makes the most sense if a dense piece of India is down in the mantle, dragging the rest of the continent down with it. That may not sound so weird, but continents are buoyant; they're supposed to float, not sink. All the subduction that is heard about all over the world is dense ocean crust sinking underneath continents.

The only Constant about Climate is Change. Significant climatic Variation can occur at decadal scale and has doneso through geological times. "Global Warming Benefits Outweigh Harms" For the past decade, according to highly accurate measurements taken from satellites, there has been no global warming. Even though atmospheric CO_2 has continued to accumulate, up about 4 percent in the past 10 years–the global mean temperature has remained flat. That should raise obvious questions about CO_2 being the cause of climate change. Orbiting satellites gather temperature readings around the globe, accurate to 0.1 degree Celsius. Warming in the upper atmosphere should occur before any surface warming effect, but NASA's data show that has not been happening. Interestingly, in the 18 years those satellites have been recording global temperatures, they actually have shown a slight decrease in average temperatures.

Other inconvenient facts ignored by the activists: Carbon dioxide is a non-polluting gas that is essential for plant photosynthesis. Higher concentrations of CO_2 in the atmosphere produce bigger crop harvests and larger and healthier forests–results environmentalists used to like. There are legitimate reasons to restrict emissions of pollutants into the atmosphere. Recycling makes sense and protecting the environment is good for everyone. But we should not fool ourselves into thinking we can change the temperature of the Earth by doing these things.

Several hypotheses have emerged to explain the global warming phenomenon. Among these theories, Anthropogenic Global Warming (AGW) has emerged as the leading theory, which deduces that the Earth is warming due to human activity and that as a result humans will eventually destroy the planet. The AGW theory has permeated nearly every facet of modern society. Over the past few decades, scientists, politicians, marketers, and Earth lovers alike, for various reasons, have employed propaganda techniques to promote the Anthropogenic Global Warming theory; in the process, they proactively sought to curtail the exposure of the truth in order to profit from people's apprehension of global warming. "The average temperature history of Antarctica provides no evidence of twentieth century warming." "The results of several research studies argue strongly against claims that CO_2-induced global warming would cause catastrophic disintegration of the Greenland and

Antarctic Ice Sheets. "The mean rate of global sea level rise has not accelerated.

The USA is the country in which the Environmental Protection Agency Has declared Carbon Dioxide to be a pollutant, and yet, 30 Coal-fired Power Plants have been built there since 2008. Coal fired power plants are the cheapest to operate with the longest life span. The construction wave stretches from Arizona to Illinois and South Carolina to Washington, and comes despite growing public wariness over the high environmental and social costs of fossil fuels, demonstrated by tragic mine disasters in West Virginia, the Gulf oil spill and wars in the Middle East.We have lots of coal then Why Can't We Build Them in Alberta? and in India as well. Hoping for a technological solution, the Obama administration devoted $3.4 billion in stimulus spending to foster "clean-coal" plants that can capture and store greenhouse gases. Yet new investments in traditional coal plants total at least 10 times that amount – more than $35 billion. Britain's Royal Society, one of the most venerable science academies, has amended its idiots' guide to global warming. Officially titled *Climate Change, A Summary of the Science*, the 19-page layman's document is a refreshing departure from the strident, doom-and-gloom message that has characterized most scientific statements on global warming, which have been parroted by the Al Gores of the world thusly: humans are to blame, sea levels will rise and the end of the world is fast approaching.

Myth 1: Global Temperatures are Rising at a Rapid, Unprecedented Rate

Fact

Accurate satellite, balloon and mountain top observations made over the last three decades have not shown any significant change in the long term rate of increase in global temperatures. Average ground station readings do show a mild warming of 0.6 to 0.8 cover the last 100 years, which is well within the natural variations recorded in the last millennium. The ground station network suffers from an uneven distribution across the globe; the stations are preferentially located in growing urban and industrial areas ("heat islands"), which show substantially higher readings than adjacent rural areas ("land use effects"). There has been no catastrophic warming recorded.

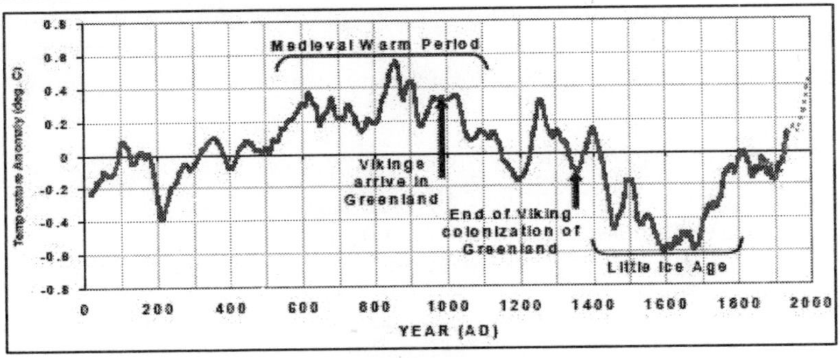

Figure 5.1

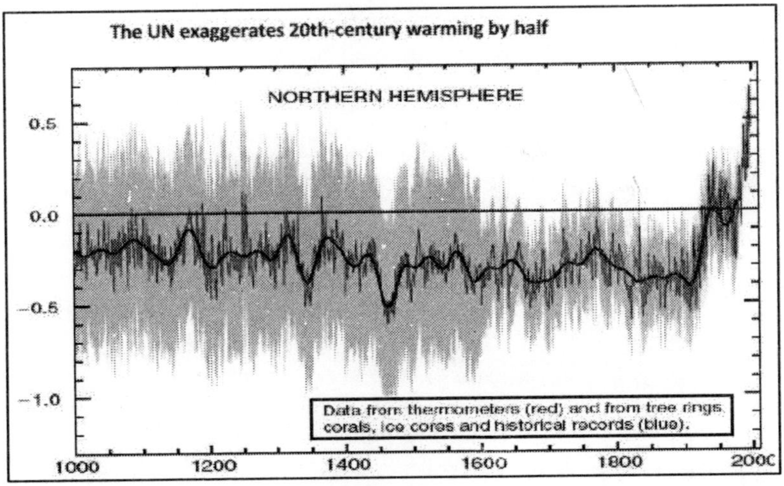

Figure 5.2

Myth 2: The "Hockey Stick" Graph Proves that the Earth has Experienced a Steady, Very Gradual Temperature Increase for 1000 Years, then Recently Began a Sudden Increase

Fact

Significant changes in climate have continually occurred throughout geologic time. For instance, the Medieval Warm Period, from around 1000 to1200 AD (when the Vikings farmed on Greenland) was followed by a period known as the Little Ice Age.

Since the end of the 17th Century the "average global temperature" has been rising at the low steady rate mentioned above; although from 1940 – 1970 temperatures actually dropped, leading to a Global Cooling scare.

The "hockey stick", a poster boy of both the UN's IPCC and Canada's Environment Department, ignores historical recorded climatic swings, and has now also been proven to be flawed and statistically unreliable as well. It is a computer construct and a faulty one at that.

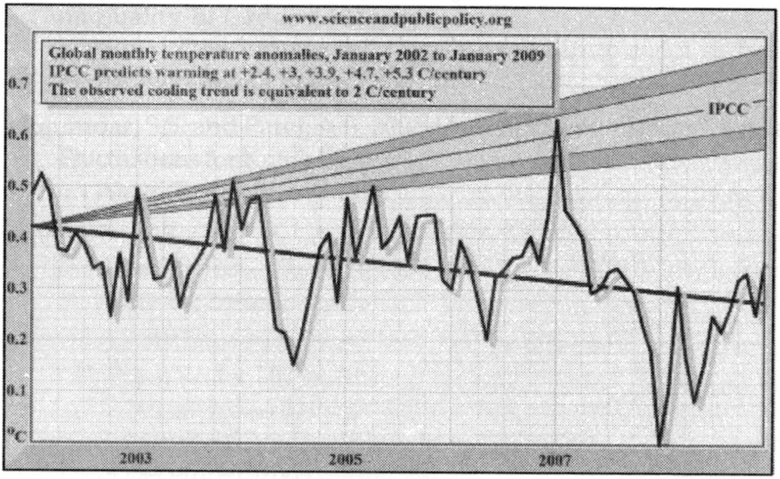

Figure 5.3

Myth 3: Human Produced Carbon Dioxide has Increased Over the Last 100 Years, Adding to the Greenhouse Effect, thus Warming the Earth

Fact

Carbon dioxide levels have indeed changed for various reasons, human and otherwise, just as they have throughout geologic time. Since the beginning of the industrial revolution, the CO_2 content of the atmosphere has increased. The RATE of growth during this period has also increased from about 0.2 per cent per year to the present rate of about 0.4 per cent per year, which growth rate has now been constant for the past 25 years. However, there is no proof that CO_2 is the main driver of global warming. As measured in ice cores dated over many thousands of years, CO_2 levels move up and down after the temperature has done so, and thus are the result of, not the cause of warming. Geological field work in recent sediments confirms this causal relationship. There is solid evidence that, as temperatures move up and down naturally and cyclically through solar radiation, orbital and galactic influences, the warming surface layers of the earth's oceans expel more CO_2 as a result.

Myth 4: CO_2 is the most Common Greenhouse Gas

Fact

Greenhouse gases form about 3 per cent of the atmosphere by volume. They consist of varying amounts, (about 97 per cent) of water vapour and clouds, with the remainder being gases like CO_2, CH_4, Ozone and N_2O, of which carbon dioxide is the largest amount. Hence, CO_2 constitutes about 0.037 per cent of the atmosphere. While the minor gases are more effective as "greenhouse agents" than water vapour and clouds, the latter are overwhelming the effect by their sheer volume and -in the end – are thought to be responsible for 60 per cent of the "Greenhouse effect". Those attributing climate change to CO_2 rarely mention this important fact.

Myth 5: Computer Models Verify that CO_2 Increases will Cause Significant Global Warming

Fact

Computer models can be made to "verify" anything by changing some of the 5 million input parameters or any of a multitude of negative and positive feedbacks in the program used. They do not "prove" anything. Also, computer models predicting global warming are incapable of properly including the effects of the sun, cosmic rays and the clouds. The sun is a major cause of temperature variation on the earth surface as its received radiation changes all the time, This happens largely in cyclical fashion. The number and the lengths in time of sunspots can be correlated very closely with average temperatures on earth, *e.g.* the Little Ice Age and the Medieval Warm Period. Varying intensity of solar heat radiation affects the surface temperature of the oceans and the currents. Warmer ocean water expels gases, some of which are CO_2. Solar radiation interferes with the cosmic ray flux, thus influencing the amount ionized nuclei which control cloud cover.

Myth 6: The UN Proved that Man–Made CO_2 Causes Global Warming

Fact

In a 1996 report by the UN on global warming, two statements were deleted from the final draft. These are following:

1. "None of the studies cited above has shown clear evidence that we can attribute the observed climate changes to increases in greenhouse gases."
2. "No study to date has positively attributed all or part of the climate change to man–made causes" To the present day there is still no scientific proof that man-made CO_2 causes significant global warming.

Myth 7: CO_2 is a Pollutant

Fact

This is absolutely not true. Nitrogen forms 80 per cent of our atmosphere. We could not live in 100 per cent nitrogen either. Carbon dioxide is no more a pollutant than nitrogen is. CO_2 is essential to life on earth. It is necessary for plant growth since increased CO_2 intake as a result of increased atmospheric concentration causes

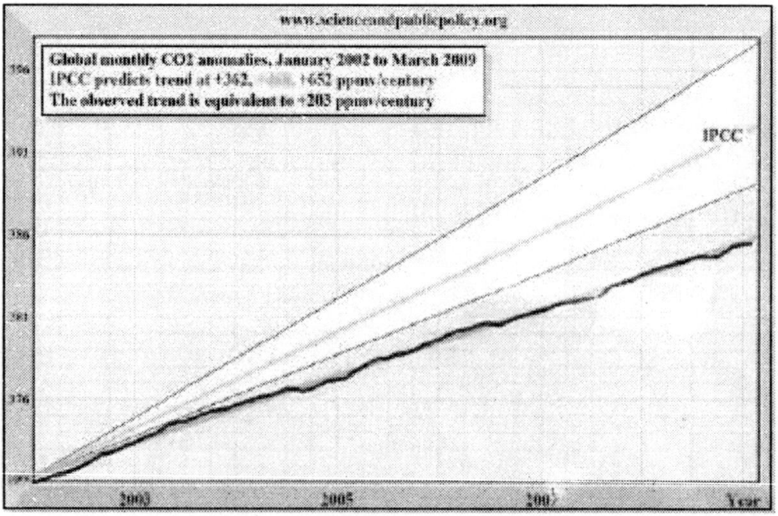

Figure 5.4

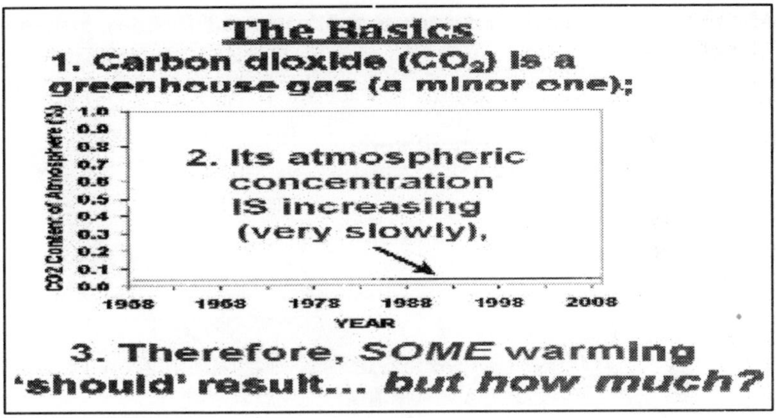

Figure 5.5

many trees and other plants to grow more vigorously. Unfortunately, the Canadian Government has included CO_2 with a number of truly toxic and noxious substances listed by the Environmental Protection Act, only as their means to politically control it.

Myth 8: Global Warming will Cause more Storms and Other Weather Extremes

Fact

There is no scientific or statistical evidence whatsoever that supports such claims on a global scale. Regional variations may occur. Growing insurance and infrastructure repair costs, particularly in coastal areas, are sometimes claimed to be the result of increasing frequency and severity of storms, whereas in reality they are a function of increasing population density, escalating development value, and ever more media reporting.

Myth 9: Receding Glaciers and the Calving of Ice Shelves are Proof of Global Warming

Fact

Glaciers have been receding and growing cyclically for hundreds of years. Recent glacier melting is a consequence of coming out of the very cool period of the Little Ice Age. Ice shelves have been breaking off for centuries. Scientists know of at least 33 periods of glaciers growing and then retreating. It's normal. Besides, glacier's health is dependent as much on precipitation as on temperature.

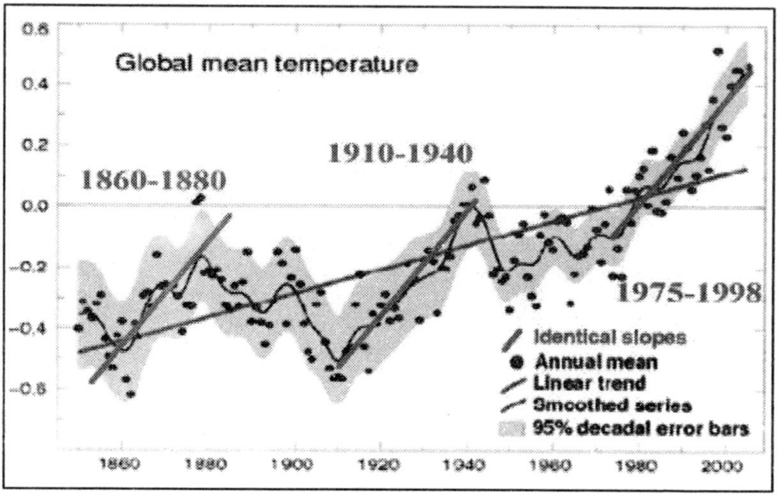

Figure 5.6: The Truth: No Anthropogenic Signal in Global Temperature

The UN's bogus graph is just one of many notorious instances of outright fabrication and data tampering in the supposedly "authoritative" reports of this apparently "august" body. This audience will not need to be told about the "hockey-stick" graph by which, six times, and in large scale, and in full color, the UN's 2001 report falsely tried to abolish the medieval warm period, pretending instead that today's temperatures are warmer than for 1000 years.

Myth 10: The Earth's Poles are Warming; Polar Ice Caps are Breaking Up and Melting and the Sea Level Rising

Warmer than the Present

Fact

The earth is variable. The western Arctic may be getting somewhat warmer, due to unrelated cyclic events in the Pacific Ocean, but the Eastern Arctic and Greenland are getting colder. The small Palmer Peninsula of Antarctica is getting warmer, while the main Antarctic continent is actually cooling. Ice thicknesses are increasing both on Greenland and in Antarctica. Sea level monitoring in the Pacific (Tuvalu) and Indian Oceans (Maldives) has shown no sign of any sea level rise.

Global Warming Propagandist Slapped Down William Connolley, through his administrative position, has for years kept dissenting views on global warming out of *Wikipedia*, allowing only those that promoted the view that global warming is a threat to mankind. Mr. Connolley's career as a propagandist for AGW has been stopped, following a unanimous verdict through an arbitration proceeding conducted by *Wikipedia*. He has been barred from participating in any article, discussion or forum dealing with global warming.

The Figure 5.2 has been tampered with not only on the left-hand side, where the medieval warm period has been artificially erased, but also on theright, where the 20th-century warming is overstated by very nearly 50 per cent.On Dr. Craig Idso's medieval-warm-period database there are now papers by more than 700 scientists from 400 institutions in 40 countries providing evidence that the warm period was real.

This graph of global temperatures (°C) over 150 years appears in the UN's 2007 climate assessment; in a lecture by the chairman of its climate-science panel, a railway engineer; in a technical document

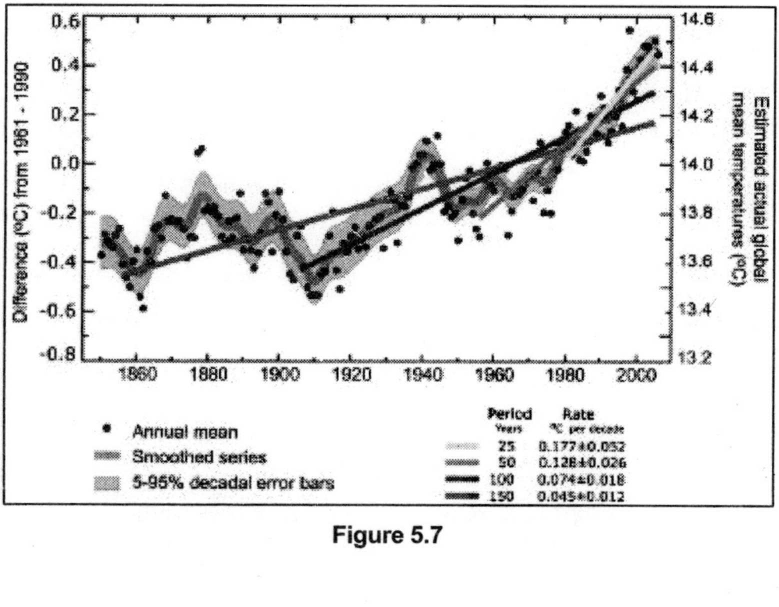

Figure 5.7

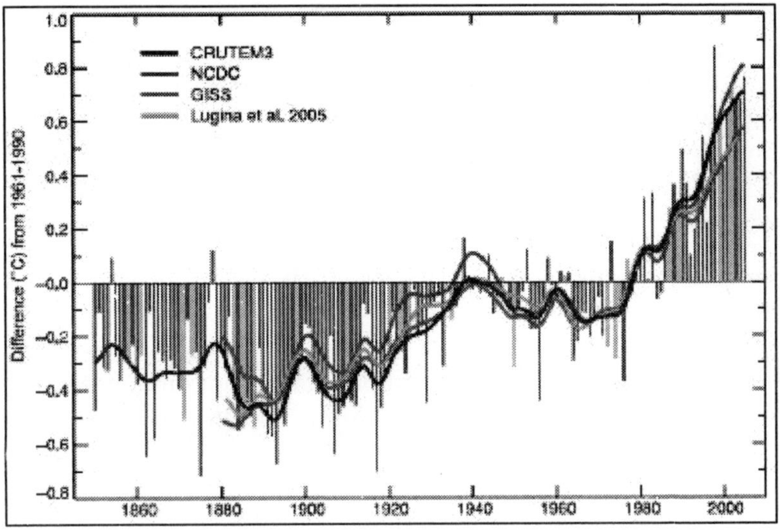

Figure 5.8: Annual Anomalies of Global Land-Surface Air Temperature (°C), 1850 to 2005, Relative to the 1961 to 1990 mean for CRUTEM3 Updated from Brohan *et al.* (2006). The smooth curve show decadal variations. The black curve from CRUTEM3 is compared with those from NCDC (Smith and Reynolds, 2005; blue), GISS (Hansen *et al.*, 2001; red) and Lugina *et al.* (2005; green)

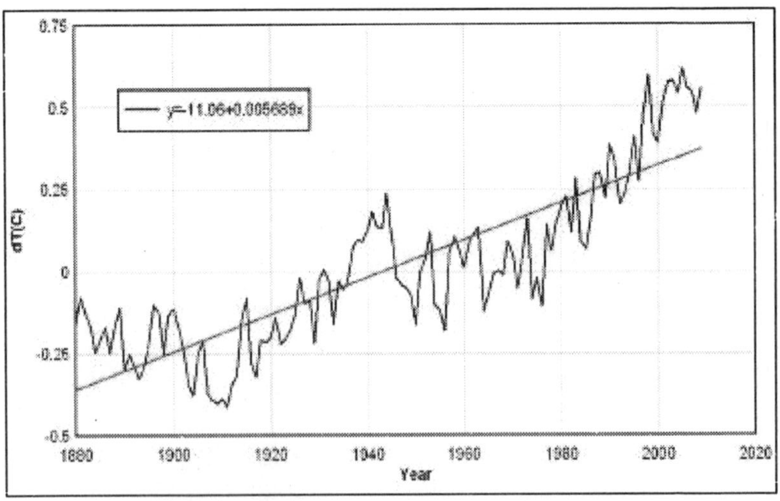

Figure 5.9: NOAA Annual Departures from Global Record Mean with Linear Trend

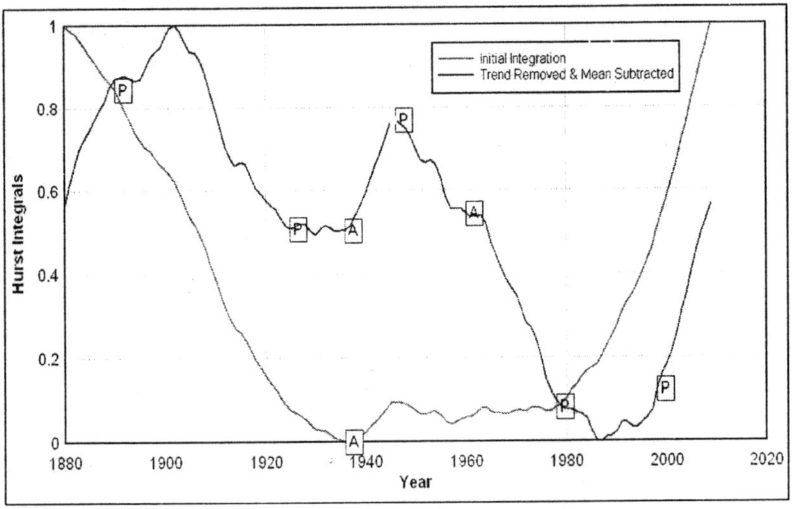

Figure 5.10: Impact of the NOAA Annual Temperature Departure Data

P: Transitions in PDO: 1890, 1925, 1976 and 1997

A: Transitions in NAO: 1935 and 1960

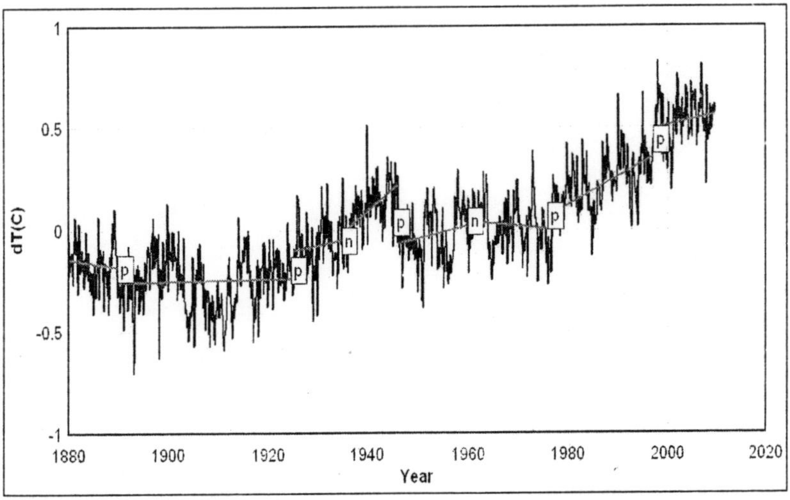

Figure 5.11: Global Monthly Temperature Deviations Trends between Ocean Circulation Transitions

P: PDO Transitions: 1890, 1925, 1946, 1976, 1997

n: NAO Transitions: 1935 and 1960

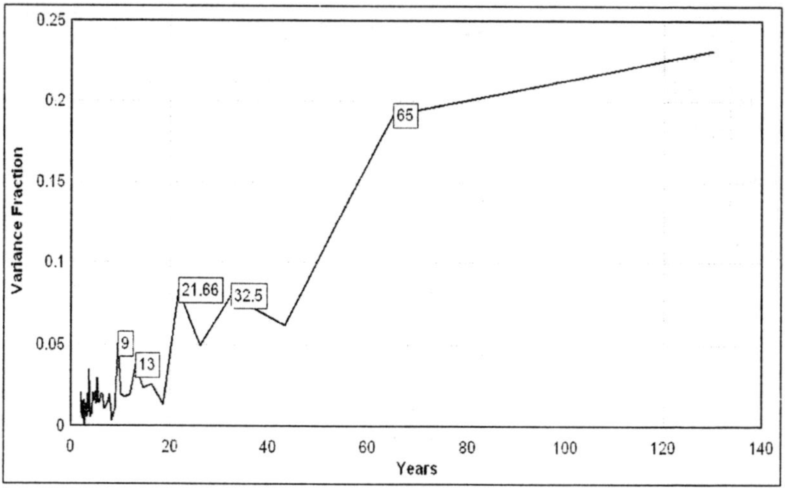

Figure 5.12: NOAA Annual Departures from Global Record Mean Fast Fourier Transform

The 2nd and 3rd Harmonics of 10.83 years and 21.66 and 32.5 years. The 6th Harmonic of 10.83 years is 65 years

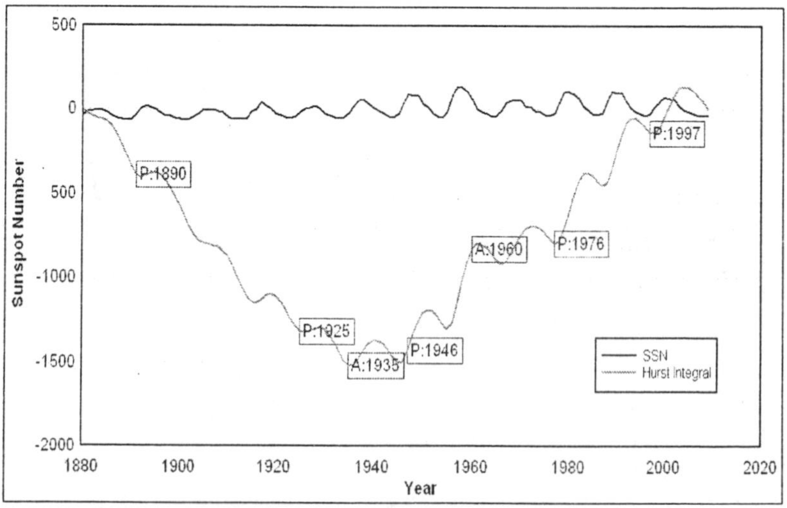

Figure 5.13: Sunspot Numers

Royal Belgian Solar Observatory: SIDC

P: PDO Transitions; A: NAO Transitions

Inflections in the Integral of the Sunspot Record are Correlated with Regime Transitions in the North Atlantic and Pacific Oceans

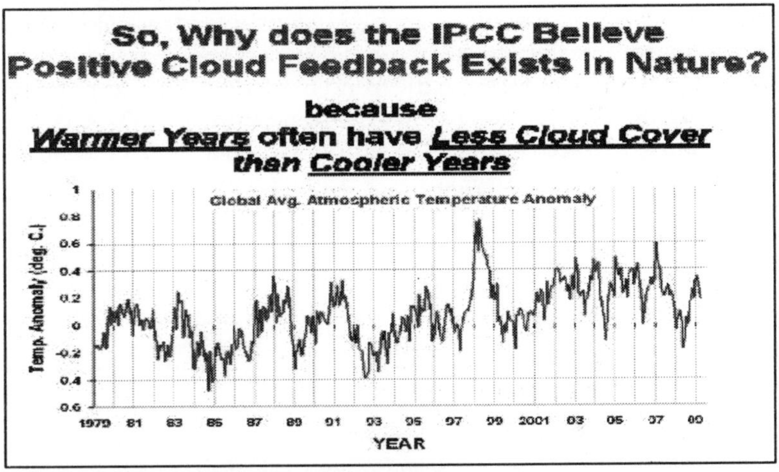

Figure 5.14

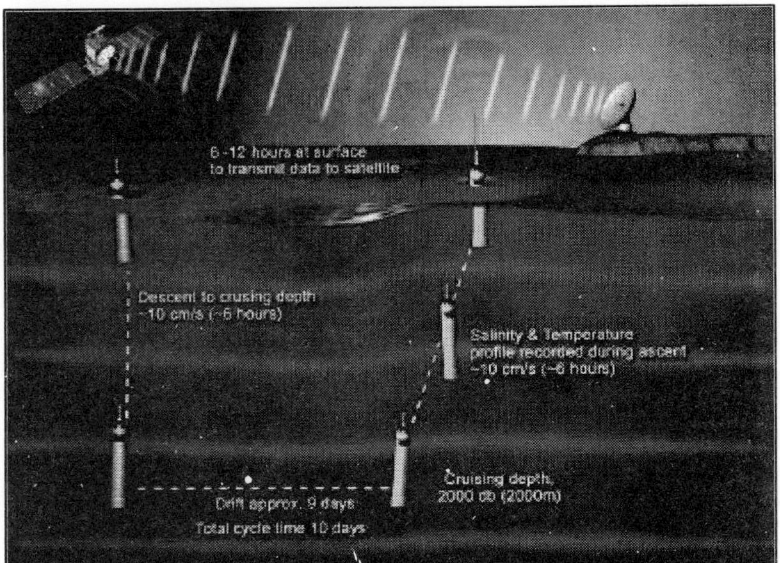

Figure 5.15: Argo Network has Over 3,000 Floats Measuring Temperature in all of the Oceans

supporting a decision by the US EPA to regulate CO_2 as a pollutant; and in a recent slide-show by a former director at Woods Hole. The "railway lines" on the graph purport to demonstrate that the rate of "global warming" is itself increasing. They are a flagrant statistical abuse. Stochastic data-curves are highly sensitive to the selection of endpoints for linear-regression trends. For instance, the trend from 1905-1945 is twice as steep as the trend from 1905-2005, apparently reversing the UN's bogus result. The trend from 1993-2009 is +1.5 °C/century; 1997-2009, +0.5 °C/century; 2001-2009, –1.5 °C/century; 2005-2009, 5.5 °C/century. We're heading for an ice age!

References

Hurst, H.E. 1951. The Long-Term Storage Capacity of Reservoirs, Transactions: *American Society of Civil Engineers*. 116:770-799.

Mantau, N.J. and Hare, S.R. 2002. The Pacific Decadal Oscillation. *Journal of Oceanography*. 58:35-44.

Outcalt, S.I., Hinkel, K.M., Meyer, E. and Brazel, A.J. 1997. Application of Hurst Rescaling to Serial Geophysical Data. *Geographical Analysis*. 29(1):72-87.

Tome, A.R. and Miranda, P.M.A. 2004. Piecewise linear fitting of trend changing points of climate parameters. *Geophysical Research Letters*. 31:L02207.

Science and Technology Committee - Eighth Report: The disclosure of climate data from the Climatic Research Unit at the University of East Anglia 2 Report by Lord Oxburgh's Science Assessment Panel.

Part-II

WATER POLLUTION: A THREAT TO LIFE

Chapter 6

Groundwater Pollution: A Threat to the Last Hope on the Blue Planet

Zeba Khanam and Vir Singh

Water is one of the most indispensable resources and is the elixir of life. Water constitutes about 70 per cent of the body weight of almost all living organisms. Life is not possible on this planet without water. It acts as a media for both chemical and biochemical reactions and also as an internal and external medium for several organisms (Rajankar *et al.*, 2009).

Water covers 70.9 per cent of the Earth's surface. On Earth, it is found mostly in oceans and other large water bodies, with 1.6 per cent of water below ground in aquifers and 0.001 per cent in the air as vapor clouds (formed of solid and liquid water particles suspended in air), and precipitation. About 97 per cent of the earth's water supply is in the oceans, where the high salt content does not permit its use for human consumption. About 2 per cent of the water resources are locked in the polar icecaps and glaciers, while only about 1 per cent is available as fresh water (surface water-rivers, lakes, streams, and groundwater) for human consumption and other use. A very small amount of the Earth's water is contained within biological bodies and manufactured products (De, 2006).

Fresh water can be obtained from various sources like streams, lakes, rivers, ponds, rain, springs, and wells. Water is known to contain a large number of chemical elements (Hutchinson, 1957) which determine its quality for various domestic, agricultural and industrial purposes. Water quality involves the physical, chemical and biological characteristics of water. It depends on the location of the source and the state of environmental protection in a given area. Therefore, the quality and the nature of water are determined by physical and chemical analysis (Voznoya, 1983).

Before water can be described as potable, it has to compile with certain physical, chemical and microbiological standards, which are designed to ensure that the water is potable and safe for drinking (Tebutt, 1983). Potable water is defined as water that is free from diseases producing microorganisms and chemical substances deleterious to health (Ihekoronye and Ngoddy, 1985). Water fit for human consumption is called drinking water or potable water.

The National Water Policy (1987) states that water is a prime natural resource, basic human need, and precious national asset. It gives special attention to drinking water for both humans and animals over its other uses. The policy calls for control of the exploitation of groundwater through regulation and an integrated and coordinated development of surface- and ground-water. The central government has identified strategies for meeting drinking water needs and micro-watershed management and conducted pilot projects in different regions in the country. Even so, India is facing a freshwater crisis.

The Groundwater

Groundwater is an important source of drinking water throughout the world. The water that is located beneath the ground surface in soil pore spaces and in the fractures of rock formations is the groundwater. A unit of rock or an unconsolidated deposit is called an aquifer when it can yield a usable quantity of water. The depth at which soil pore spaces or fractures and voids in rock become completely saturated with water is called the water table. Groundwater is recharged from, and eventually flows to, the surface naturally; natural discharge often occurs at springs and seeps and can form oases or wetlands. Groundwater is also often withdrawn for agricultural, municipal and industrial use by constructing and operating extraction wells. The study of the distribution and movement of groundwater is hydrogeology, also called groundwater hydrology.

Groundwater makes up about 20 per cent of the world's fresh water supply, which is about 0.61 per cent of the entire world's water, including oceans and permanent ice. Global groundwater storage is roughly equal to the total amount of freshwater stored in the snow and ice pack, including the north and south poles. Groundwater is naturally replenished by surface water from precipitation, streams, and rivers when this recharge reaches the water table. About 95 per cent of rural population living in India depends on groundwater for domestic use (Moharir *et al.*, 2002). Groundwater is highly valued because of certain properties not possessed by surface water (Goel, 2000).

The endless circulation of water as it moves in its various phases through the atmosphere, to the Earth, over and through the land, to the ocean, and back to the atmosphere is known as the hydrologic cycle. This cycle is powered by the Sun; through phase changes of water (*i.e.* evaporation and condensation) involving storage and release of latent heat, it affects the global circulation of both the atmosphere and oceans, and hence is instrumental in shaping weather and climate. The efficiency of water as a solvent makes geochemistry an intimate part of the hydrologic cycle; all water-soluble elements follow this cycle at least partially. Thus, the hydrologic cycle is the integrating process for the fluxes of water, energy, and the chemical elements. This cycle is the foundation of hydrological science and occurs over a wide range of space and time scales.

Water enters the hydrologic system as precipitation, in the form of rainfall or snowmelt. It leaves the system as stream-flow or runoff, and as evapo-transpiration, a combination of evaporation from open bodies of water, evaporation from soil surfaces, and transpiration from the soil by plants. Precipitation is delivered to streams on the land surface as overland flow to tributary channels, and in the subsurface as interflow or lateral subsurface flow and base-flow following infiltration into the soil. A portion of the infiltrated water enters the groundwater or aquifer system by passing through the vadose or unsaturated zone, and it exits to the atmosphere, surface water, or to plants. The flow-lines deliver groundwater from the highlands towards the valleys, or from the recharge areas to the discharge areas. In a recharge area there is a component to the direction of groundwater flow that is downward. Groundwater recharge is the entry to the saturated zone of water made available at the water table surface. Conversely, in a discharge area there is a component to the direction of groundwater flow that is upward. Groundwater discharge is the removal of water from the saturated zone across the water table surface. The patterns of groundwater flow from the recharge to the discharge areas form groundwater flow systems, which constitute the framework for understanding recharge processes.

The quality of groundwater is the resultant of all the processes and reactions that act on the water from the moment it condensed in the atmosphere to the time it is discharged by a well or spring and varies from place to place and with the depth of the water table. It depends upon the nature of rock formation, physiography, soils, environment, recharge and discharge conditions in the area. As groundwater moves underground it tends to develop a chemical equilibrium by chemical reactions. Being an important and integral part of the hydrological cycle, its availability depends on the rainfall and recharge conditions. Groundwater has unique features, which render it particularly suitable for public water supply. It has excellent natural quality, usually free from pathogens, colour and turbidity that can be consumed directly without treatment. Till recently it had been considered a dependable source of uncontaminated water (Naeem *et al.*, 2004).

Groundwater Pollution

Water contamination can be scientifically defined as "the addition of any substance to water, changing water's physical and chemical characteristics in any way, which interferes with its use for legitimate purposes".

Normally, water is never pure in a chemical sense. It contains impurities of various kinds in the form of dissolved as well as in suspended form. It includes dissolved gases (like H_2O, CO_2, NH_3, N_2), dissolved minerals (like Ca, Mg, Na salt), suspended matter (clay, silt, and salt) and even microbes. These are natural impurities in water, which are derived from atmosphere, catchment areas and soils. They are present in very low amount and are potable in water.

Water pollutants are mainly of three types:

1. Inorganic pollutants
2. Organic pollutants
3. Biological pollutants

Under inorganic pollutants, the best examples are heavy metals. Others are radio nuclides, phosphates, mostly from detergents, nitrates generally from sewage and fertilizers. Organic pollutants are mainly generated by pesticides, herbicides and common households and industrial material which are common in use. Common examples of biological pollutants are bacteria, virus, fungi which serve as agent of many contagious diseases like diarrhea, cholera etc.

The main source of fresh water pollution can be attributed to discharge of untreated wastes, dumping of industrial effluents and agricultural run-off. Industrial growth, urbanization and the increasing use of synthetic organic substances have serious and adverse impacts on fresh water bodies. The freshwater crisis often provoked by the improper management of water resources and environmental degradation, which has led to a lack of access to safe water supply to millions of people.

A groundwater pollutant is any substance that, when it reaches an aquifer, makes the water unclean or otherwise unsuitable for a particular purpose. Sometimes the substance is a manufactured chemical, but just as often it might be microbial contamination. Contamination also can occur from naturally occurring mineral and metallic deposits in rock and soil.

For many years, people believed that the soil and sediment layers deposited above an aquifer acted as a natural filter that kept many unnatural pollutants from the surface from infiltrating down to groundwater. By the 1970s, however, it became widely understood that those soil layers often did not adequately protect aquifers. Despite this realization, a significant amount of contamination already had been released to the nation's soil and groundwater. Scientists have since realized that once an aquifer becomes polluted, it may become unusable for decades, and is often impossible to clean up quickly and inexpensively.

Groundwater pollution caused by human activities usually falls into one of two categories:

1. Point-source pollution and
2. Nonpoint-source pollution.

Point-source pollution refers to contamination originating from a single tank, disposal site, or facility. Industrial waste disposal sites, accidental spills, leaking gasoline storage tanks, and dumps or landfills are examples of point sources.

Chemicals used in agriculture, such as fertilizers, pesticides and herbicides are examples of nonpoint-source pollution because they are spread out across wide areas. Similarly, runoff from urban areas is a nonpoint source of pollution.

Microbial Contamination of Groundwater

Groundwater near the land surface usually teems with microbial life. Bacteria, which are microscopic single-celled organisms that lack a true nucleus and normally have a cell wall, are far more numerous than any other organism in the soil and groundwater. Near the surface where plant roots are abundant, there may be 100 million to 1 billion bacteria per gram of dry soil. These values decrease dramatically with soil depth below the root zone, with densities depending on the amount of nutrients (food) and water available. Although concentrations of microbes below the root zone are lower than in the root zone itself, as many as 10 to 100 million bacteria per gram of aquifer material may be present.

Microbes below Earth's surface must compete vigorously with each other for limited food and space. Bacteria that are not adapted to this habitat cannot survive long. Normally, nutrients are scarce in this environment, and thus microbes in groundwater grow slowly. Because the microbes in groundwater and other environmental habitats are well adapted to life in these habitats, few of them have the ability or the need to infect and cause illness in humans and other animals. However, a few environmental bacteria and protozoa may cause disease in humans by chance under certain circumstances, especially in individuals with a weakened immune system or major breaks in the skin that allow microbes to enter.

The groundwater in a drinking-water well may contain a wide variety of microbes without presenting a public health risk. However, groundwater in some areas becomes contaminated by the fecal material of humans and other animals. This is a cause for concern because fecal material may contain pathogenic (disease-causing) microbes that can infect the intestinal tract of humans. Fecal pathogens may be bacterial, viral, or protozoan.

Water containing fecal material may seep into the groundwater from the land surface or from underground sources of contamination. Major surface sources include:

- Wastewater and biosolids from sewage treatment facilities that have been applied to land as a soil conditioner;
- Seepage from shallow artificial ponds (lagoons) used for processing sewage;
- Seepage from contaminated lakes and other surface-water bodies;
- Urban runoff;
- Feces from cattle and other livestock operations; and
- Improperly constructed sanitary landfills where trash and garbage are disposed.

Fecal contamination also can reach the groundwater from underground sources, such as improperly functioning septic tank systems, underground reservoirs for liquid household sewage (cesspools), or leaking underground sewer lines.

Polluted water is turbid, unpleasant, having bad smell and is unfit for drinking as well as bathing and washing purposes. Polluted water is harmful and is a vehicle for many diseases, such as cholera, dysentery, typhoid etc. Other diseases are like abdominal pain, skin lesions, dizziness, constipation, indigestion, burning sensations in chest and stomach, skin discoloration, suppression of lactation, underweight births, birth defects, child growth retardation and contaminated breast milk.

Urbanization and the unregulated growth of the population have altered the surface and subsurface terrains of the many areas. Changes in local topography and drainage system directly affect both quality and quantity of the groundwater (Vasanthavigar *et al.*, 2010).

Temporal changes in the origin and constitution of the recharged water, hydrological human factors may cause periodic changes in the groundwater quality. Water pollution not only affects water quality but also threats human health, economic development and social prosperity (Milovanovic, 2007).

River basins are highly vulnerable to pollution due to absorption and transportation of domestic, industrial and agricultural waste water; therefore it is significant to control water pollution and monitor water quality (Simeonov *et al.*, 2003).

Over burden of the population pressure, unplanned urbanization, unrestricted exploration policies and dumping of the polluted water at inappropriate place enhance the infiltration of harmful compounds to the groundwater. Studies regarding the groundwater quality analysis have been made by many authors (Gupta and Gupta, 1999; Rajasekara *et al.*, 2005; Rajan and Paneerselvam, 2005; Thakare *et al.*, 2005; Bisht *et al.*, 2007). They concluded that it is the high rate of exploration then its recharging, inappropriate dumping of solid as well as liquid wastes, lack of strict enforcement of law and loose governance are the cause of deterioration of groundwater quality (Pandey and Tiwari, 2009).

It has been estimated that once pollution enters the subsurface environment, it may remain concealed for many years, becoming dispersed over wide areas of groundwater aquifer and rendering groundwater supplies unsuitable for consumption and other uses. The rate of depletion of groundwater levels and deterioration of groundwater quality is of immediate concern in major cities and towns of the country (Khurshid *et al.*, 1997; Das *et al.*, 1998; Garode *et al.*, 1998; Kapley *et al.*, 1998; Naidu *et al.*, 1998; Jain *et al.*, 2000a,b,c; Sohani *et al.*, 2001; Jain, 2002; Jain *et al.*, 2003a,b,c; Meenakumari and Hosmani, 2003; Dhindsa *et al.*, 2004; Jain, 2004a,b,c; Jain *et al.*, 2004; Ramasubramanian *et al.*, 2004).

The typically low velocity of groundwater prevents a great deal of mixing and dilution, consequently, a contaminant plume may maintain a high concentration as it slowly moves from points of recharge to zone of discharge (Pattyjohns, 1979).

Water soluble substances that are dumped, spilled, spread or stored on the land surface eventually may infiltrate. Groundwater can also become contaminated by the disposal of fluids through wells and, in lime stone terrains, through sinkholes directly into aquifers. Likewise, infiltration of contaminated surface water has caused groundwater contamination in several places. Irrigation tends to increase the mineral content of both surface and groundwater.

The degree of severity in such cases is related to hydrologic properties of the aquifer, the type and amount of waste, disposal method and climate. Another cause of groundwater quality deterioration is excessive pumping out, which may precipitate the migration of more mineralized water from the surrounding strata to the well. Various studies (Kaul *et al.*, 1999; Khaiwal and Garg, 2006; Yadav and Lata, 2003; Meenakshi *et al.*, 2004; Prakash and Somashekar, 2006; Bishnoi and Arora, 2007; Shailaja and Jhonson, 2007) carried out in the past have reported the presence of excessive fluoride, arsenic, nitrate, sulphate, heavy metals, salinity, hardness, pesticides etc. from different parts of India. It has been reported that 77 per cent of urban population and only 31 per cent of rural population in India has access to potable water (Bishnoi and Malik, 2007).

Industrialization is the index of modernization which leads to alteration in the physical, chemical and biological properties of environment. The history of human civilization reveals that water source and civilization runs parallel (Sonawanea and Shrivastava, 2010). The era of unlimited fresh water supply is coming to end due to pollution of water sources owing to the increasing discharge of large volume of waste water and toxic nature of wastes (Mohapatra and Singh, 1998). Much of the health in developing countries is largely due to lack of the safe drinking water. According to WHO, about 600 million cases of diarrhea and 46,00,000 childhood deaths are reported per year because of contaminated water and lack of sanitation (Sahu *et al.*, 2006).

Although India has substantial freshwater resources there is an acute shortage of safe drinking water of acceptable quality (Batheja, Sinha and Seth, 2008). Most of the Indian rivers and freshwater streams are seriously polluted by textile dyeing and printing industry effluents, which includes wastes like metals, detergents, acids, alkalis, sulfates chlorides, nitrates, dissolved and suspended solids, organic and microbial impurities (Naik and Choukmath, 2002).

Our primary and foremost objective nowadays is to save water, *i.e.* Earth's blue gold, and its purity. Changes in habit can reduce water in irrigation by 50 per cent, in domestic usage by 35 per cent and water quality can also be checked (Mahapatra and Mahanty, 2005; Sinha, 1995).

Last Hope of the Blue Planet?

Surface water, including river and stream water, is already polluted to a great extent. In most of the cases, river water is unfit for consumption. The underground water remains to be the last hope to be dependent upon for healthy life. But this source of water too is gradually polluting and sooner or later would turn into a grave threat to health of the planet. The underground water sources need to be saved from getting polluted, which is not only necessary but an imperative of our life.

References

Batheja, K., Sinha, A.K. and Seth, G. 2008. *J of Indian water works Association*, 40(1): 45-49.

Bishnoi, M.S. and Malik. 2008. Ground water quality in environmentally degraded localities of Panipat city, India. *J. Environ. Biol.* 29(6): 881-886.

Bishnoi, M.S. and Arora, S. 2007. Potable groundwater quality in some villages of Haryana, India: Focus on fluoride. *J. Environ. Biol.* 28: 291-294.

Bisht, Shikha, Patra, B. A., Gupta, N. C., Arora, S., and Singh, R. A. 2007. Assessment of Drinking Water Quality of Delhi, India.Cidade de Goa, Dona Paula, Goa.

Das, S., Metha, B. C., Das, P. K., Srivastava, S. K., and Samanth, S. K. 1998. Source of high fluoride in ground water of Anugul, Dhenkenal District, Orrissa. *Pollution Research* 17(4): 385–392.

De, A.K. 2010. Enviromental Chemistry. *New Age International Publishers* 6th edition: 5.

Dhindsa, S. S., Bheel, P., and Musturia, Y. 2004. Hydrochemical study of ground water quality variation in Tonk District, Rajasthan. *Indian Journal of Environment and Ecoplanning* 8(1): 129–136.

Garode, A.M., Nanoty, V. D., and Bodhankar, M.G. 1998. Bacteriological status of drinking water in and around Chikhli Town of Buldana District, Maharashtra. *Pollution Research* 17(3): 293–294.

Goel, P.K. 2000. Water Pollution - Causes, Effects and Control, *New Age Int. (P) Ltd.*, New Delhi.

Gupta, B.K. and Gupta, R.R. 1999. Physio-chemical and biological study of drinking water in Satna, Madhya Pradesh. *Poll. Res.* 18: 523-525.

Hutchinson, G.E. 1957. A Treatise on Limnology. *Palaeography, Palaeoclimatology, Palaeoecology* 130(4): 307-321.

Ihekoronye, A.I. and Ngoddy, P.O. 1985. Integrated Food Sciences and Technology for the Tropics. *Macmillan Press London, Oxford* : 95-195.

Jain, C. K. 2002. Arsenic contamination in ground water: Indian scenario. *Indian Journal of Environmental Health* 44(3): 238–243.

Jain, C. K. 2004a. Ground water quality of District Dehradun, Uttaranchal. *Indian Journal of Environment and Ecoplanning* 8(2): 475–484.

Jain, C. K. 2004b. Irrigation water quality in District Dehradun, Uttaranchal. *Indian Journal of Environment and Ecoplanning* 8(2): 319–326.

Jain, C. K. 2004c. Trace element contamination in ground water of District Dehradun, Uttaranchal. *Indian Journal of Environment and Ecoplanning* 8(3): 597–605.

Jain, C.K., Bandyopadhyay, A. and Bhadra, A. 2010. Assessment of ground water quality for drinking purpose, District Nainital, Uttarakhand,India. *Environ Monit Assess.*166: 663-676.

Jain, C. K., Bhatia, K. K. S., and Kumar, S. R. 2000b. Ground water contamination in Greater Guwahati, Assam. *Indian Journal of Environmental Protection* 20(9): 641–648.

Jain, C.K., Bhatia, K.K.S., and Kumar, V. 2000a. Ground water quality in Sagar district, Madhya Pradesh. *Indian Journal of Environmental Health* 42(4): 151–158.

Jain, C.K., Bhatia, K.K.S., Kumar, C.P., and Purandara, B.K. 2003a. Ground water quality in Malprabha sub-basin, Karnataka. *Indian Journal of Environmental Protection* 23(3): 321–329.

Jain, C.K., Kumar, C.P., and Sharma, M.K. 2003b. Ground water quality in Ghataprabha Command Area, Karnataka. *Indian Journal of Environment and Ecoplanning* 7(2): 251–262.

Jain, C.K., Kumar, C.P., and Sharma, M.K. 2003c. Irrigation water quality in Ghataprabha Command Area, Karnataka. *Indian Journal of Environment and Ecoplanning* 7(2): 409–420.

Jain, C.K., Kumar, S., and Rao, Y.R.S. 2004. Trace element contamination in a coastal aquifer of Andhra Pradesh. *Pollution Research* 23(1):13–23.

Jain, C.K., Sharma, M.K., Bhatia, K.K.S., and Seth, S.M. 2000c. Ground water pollution–Endemic of fluorosis. *Pollution Research* 19(4): 505–509.

Kapley, R.D., Patode, H.S. and Panaskar, D.B. 1998. Ground water quality in an industrial area of Tuppa, Nanded, Maharashtra. *Pollution Research,* 17(3): 251–255.

Kaul, S.N., Mahajan, A.U. and Nandy, T. 1999. Water and waste water treatment, recycle and reuse. *J. Ind. Assoc. Environ. Manage.* 26: 74-90.

Khaiwal, R. and Garg V.K. 2006. Distribution of fluoride in groundwater and its suitability assessment for drinking purpose. *Inte. J. Environ. Hlth. Res.* 16: 163-166.

Khurshid, S., Zaheeruddin, and Shbeer, M.U. 1997. Degradation of water quality due to heavy metal pollution in Faridabad District, Haryana, India. *Pollution Research* 16(1): 41–43.

Mahapatra, M.K. and Mohanty, A.K. 2005. Ground water pollution in Subarnapur and Nuapada district of Orrisa,India. *Poll. Res.* 24(4): 863-865.

Meenakshi, Garg, V.K., Kavita, Renuka and Malik A. 2004. Groundwater quality in some villages of Haryana, India: Focus on fluoride and fluorosis. *J. Hazrdous Matr.* 106: 85-97.

Meenakumari, H.R., and Hosmani, S.P. 2003. Physicochemical and biological quality of ground water in Mysore City, Karnataka. *Indian Journal of Environment and Ecoplanning* 7(1): 79–82.

Milovanovic, M. 2007. Water quality assessment and determination of pollution sources along the Axios Vardar River, *Southeastern Europe. Desalination,* 213: 159-173.

Mohapatra, U.K., Singh, B.C. 1998. Groundwater quality assessment nearer to the dye user Industry. *Ind J Env. Pollu.* 18(7): 532-535.

Moharir, A., Ramteke, D.S., Moghe, C.A., Waste, S.R. and Sarin, R. 2002. Surface and Ground water quality assessment in Bina region. *Ind. J. Environ. Protec.* 22(9): 961-969.

Naeem, M., Khan, K., Rehman, S., and Javed, I. 2004. Environmental Assessment of Ground Water Quality of Lahore Area, Punjab, Pakistan. *J. Applied Sci.* 7: 41-46.

Naidu, M.N., Srinivas, N., and Prasada, R. P.V. 1998. Urbanwater quality in North Coastal Andhra Pradesh. *Indian Journal of Environmental Protection* 18(4): 273–277.

Naik, S.O. and Choukmath, A. 2002. Effective management of textile effluents-manmade textile of India, 443-445.

Pandey, S. K., Tiwari, S. 2009. Physico-chemical analysis of ground water of selected area of Ghazipur city. *Nature and Science* 7(1): 17-20.

Pattyjohns, W.A. 1979. Groundwater pollution - An imminent disaster. *Groundwater* 17: 27-51.

Prakash, K.L. and Somashekar, R.K. 2006. Groundwater quality–assessment on Anekal Taluk, Bangalore Urban district, India. *J. Environ. Biology*. 27(4): 633-700.

Rajan, M.R. and Paneerselvam, I. 2005. Evaluation of drinking water quality in Dindigul city, Tamil Nadu. *Indian J. Environ. and Ecoplan.* 10(3): 771-776.

Rajankar, P.N., Gulhane, S.R., Tambekar, D.H., Ramteke, D.S. and Wate, S.R. 2009. Water Quality Assessment of Groundwater Resources in Nagpur Region (India) Based on WQI. *Journal of Chemistry* 6(3): 905-908.

Ramasubramanian, V., Jeyaprakash, R., Mallika R., Ramasubbu, R., and Mariappan, V. 2004. Analysis of physico-chemical characteristics of ground water quality and quality index in and around Sivakasi Town. *Indian Journal of Environment and Ecoplanning* 8(1): 171–176.

Sahu, A.K., Pandey, A.B., and Salam, R. 2006. Ground Water Quality Assessment Nearer to the Dye user Industry. *Pollution Research* 25(2): 333-335.

Shah, B.A. 2008. Role of Quaternary stratigraphy on arsenic-contaminated groundwater from parts of Middle Ganga Plain, UP–Bihar, India. *Environ. Geol.* 53: 1553–1561.

Shailaja, K. and Johnson, M.E.C. 2007. Fluorides in groundwater and its impact on health. *J. Environ. Biol.* 28: 331-332.

Simeonov, V., Stratis J.A., Samara, C., Zachariadis, D. V. and Anthemidis, A. 2003. Assessment of the surface water quality in Northern Greece. *Water Res.*, 37: 4119-4124.

Sinha, D.K. and Srivastava, A.K. 1995. Physico-chemical characteristics of river Sai at Raebareilly. *Indian J. Environ. Hlth.* 37(3): 205-210.

Sohani, D., Pande, S., and Srivastava, V.S. 2001. Ground water quality at Tribal Town: Nandurbar (Maharashtra). *Indian Journal of Environment and Ecoplanning* 5(2): 475–479.

Sonawane, G.H. and Shrivastava, V.S. 2010. Ground Water Quality Assessment Nearer to the Dye user Industry. *Archives of Applied Science Research* 2 (6): 126-130.

Tebutt, T.H.Y. 1983. Principles of Quality Control. *Pergamon*, England: 235.

Thakare, S.B., Parvate, A.V. and Rao, M. 2005. Analysis of fluoride in the ground water of Akola district. *Indian J. Environ. and Ecoplan.* 10(3): 657-661.

Vasanthavigar, M., Srinivasamoorthy K., Vijayaragavan K., Ganthi, R., Chidambaram S., Anandhan P., Manivannan R. and Vasudevan S. 2010. Application of water quality index for groundwater quality assessment: Thirumanimuttar sub-basin, Tamilnadu, India. *Environ Monitoring Assess.* DOI 10.1007/s10661-009-1302-1.

Voznoya, F.N. 1983. Chemistry of water and Microbiology. *Mir publishers* 2

Yadav, J.P. and Lata. 2003. Urinary fluoride levels and prevalence of dental fluorosis in children of Jhajjar district, Haryana. *Ind. J. Med. Sci.* 57: 394-399.

Chapter 7

Assessment of Pesticide Levels in Human System Due to Environmental Contamination from North-East India

Kumkum Mishra and Ramesh C. Sharma

Organochlorine pesticides have been of tremendous benefit to man and his environment but their inadvertent use has caused considerable harm to human health. Exposure of human to these hazardous chemicals occurs directly in the fields and indirectly due to consumption of contaminated diet or by inhalation or by dermal contact. Despite the proliferation of different types of pesticides, organochlorines such as HCH and DDT still account for two third of the total consumption in the country because of their low cost and versatility in action again various pests. Since Assam, a state in north-eastern part of India is an endemic area with perennial transmission where mosquito borne diseases are serious health problems; these pesticides are still being used in huge quantities in disease vector control. Therefore it will be interesting to determine the residue levels of DDT and HCH in human blood samples from district Nagaon (high malaria incidence and high DDT consumption) and Dibrugarh (low malaria incidence and low DDT consumption) of Assam. A total of 331 human blood samples have been collected from these two districts dividing each district into different locations. The concentration of selected persistent organochlorine pollutants (o,p′-DDE, p,p′-DDE, p,p′-DDD, o,p′-DDT, p,p′-DDT, α-HCH, β-HCH, γ-HCH and δ-HCH) was measured with gas chromatography-electron capture detector. All human blood samples were found to be contaminated with different levels of DDT and HCH residues. The results demonstrated that the mean levels of total DDT and HCH were 742.61μg/L and 627.10 μg/L for district Nagaon while 416.86 μg/L and 348.11 μg/L for district Dibrugarh. The difference of Total HCH and Total DDT between the districts is found to be highly significant with P=0.000. Among DDT metabolites, same trend is observed for both the districts as p,p-DDT is the principal component contributed 41 per cent for Nagaon and 58 per cent for Dibrugarh. In case of HCH residues, for Nagaon, β-HCH is the predominant contaminant comprised more than 34 per cent while for Dibrugarh, it is α-HCH that is contributing 43 per cent of total HCH concentration.

Introduction

Persistent organic pollutants (POPs) have been major environmental issue, drawing most scientific and public attentions as they are characterized by their persistent, fat-seeking and endocrine disruptive nature (Hansen, 1998; Hermann, 1993). Organochlorine pesticides (OCPs) were extensively used to increase agricultural products by preventing losses due to pest. The public health department also uses these chemicals to control various vectors, which spread diseases like malaria, plague, etc. The organochlorine compounds are considered to be persistent pollutant. These compounds cause serious problem because they persist in the environment for a long time. These toxicants enter the human body through the food chain and cause serious health problems (John *et al.*, 2001).

Among the organochlorine pesticides, DDT and HCH excel in their broad spectrum toxicity and residual activity. Developed countries have banned the usage of DDT since 1975, due to its inherent chemical stability, persistence, high liposolubility and bioaccumulation. Despite these undesirable properties, developing countries of Asia are still using these pesticides due to cost-benefit efficiency and broad spectrum toxicity (Gupta, 2004; Lallas, 2001). DDT and HCH are banned in India for agriculture but are still used for control of vectors in public health. HCH was banned for use in agriculture in 1997 (Mukherjee and Gopal, 2003).

DDT and HCH are well known organochlorine pesticides as both were extensively used in pest control programme in agriculture and public health. In India, DDT has been totally banned for agricultural use. India has sought exemption under Stockholm Convention for use of 10,000 tons of DDT for its restricted use in public health sector for mosquito control (Lallas, 2001). Though the usage of technical HCH is finally banned, the government of India is encouraging its replacement with Lindane (γ-HCH), which has all the hazardous properties of HCH.

Different groups of population are exposed in different ways and in varying degrees due to occupational and non occupational exposures. The occupational exposures could be during manufacture and formulation of pesticides in industrial settings and their distribution in field condition during application of pesticides. The non occupational exposures or indirect toxic effects may be due to pollution of the ecosystem through food chain. The organochlorine pesticides are potentially toxic, highly persistent and readily accumulate in human body tissues causing health hazards (Ceron *et. al.*, 1995).

Organochlorine pesticide residues are considered endocrine disrupters and are also suspected as carcinogens (Amaral Mendes, 2002; Brody and Rudel, 2003; Calle *et al.*, 2002; Charlier and Plomteux, 2002; Gammon *et al.*, 2002). β-HCH and p,p'-DDT showing estrogenic effects and p,p'-DDE being reported as an androgen antagonist (Kunisue *et al.*, 2004). Several organochlorine pesticides can cause adverse effects in the male reproductive system in animals (Edwards *et al.*, 2006; Vos et. al., 2000). Recent reports have suggested an association between exposure to pesticides and various types of human cancer (Amaral Mendes, 2002; Wolff *et al.*, 2000). Presence of OCPs has also been related to increased risk of environmental-induced breast cancer (Ahmed *et al.*, 2002; Frias *et al.*, 2001; Snedeker, 2001; Wolff *et al.*, 2000).

Organochlorine compounds such as polychlorobiphenyl (PCBs), hexachlorohexanes (HCHs), hexachlorobenzene (HCB) and DDTs are ubiquitous contaminants on our planet. Once release in to the environment they may be transported over long distance (Wania and Mackay, 1996) and be incorporated in too many biogeochemical cycles without undergoing important degradation.

The aim of this study was to determine the levels of OC-DDTs and HCHs in human blood from the population of selected districts of Assam, India which are located in an area of intense application of pesticides in public health for malaria control as well as for agriculture, and for that, with higher environmental exposure. We have tried to collect useful data on contamination levels in North-east region of India that has not been yet reported. Additionally, the source of interest could be to compare the results for the total DDT and HCH body burden with the results obtained from HCH isomers and DDT metabolites, possible differences between age, gender and higher blood organochlorine levels.

Material and Methods

Study Area

The present study was performed in two districts of state of Assam (latitude 24° 44´ to 27° 45´ N and longitude 89° 41´ to 96° 02´ E) - One with low malaria incidence and low pesticide application (District Dibrugarh) and the other with high malaria incidence and high pesticide application (District Nagaon) (Figure 7.1).

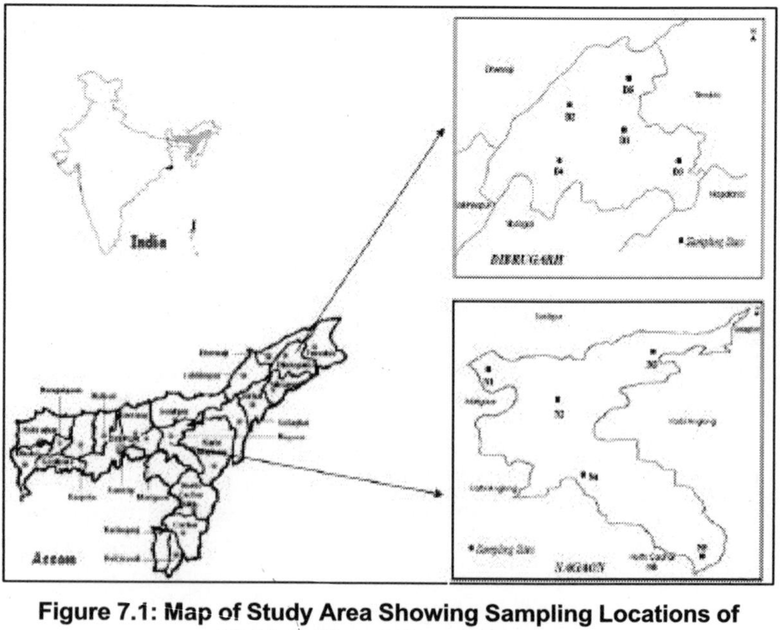

Figure 7.1: Map of Study Area Showing Sampling Locations of District Dibrugarh and Nagaon

Study Group and Sample Collection

A total of 331 samples (162 from Nagaon district and 169 from Dibrugarh district) were collected from different locations representing the whole district including one urban area (D2 and N2) from both the districts out of which 152 were males and 179 were females. The female's ages were between 9-76 years while males were between 6-80 years. Sample size have been calculated using Lot Quality Assurance sampling method (Tawfik *et al.*, 2001) with a confidence limit of 95 per cent from each side.

Blood samples were collected and processed as per the method reported by Dua *et al.*, 1996. Approximately 100 ml of blood was collected with the help of heparinized marked capillary glass tube of 100ml capacity (Drumand Scientific Co. USA) on filter paper by pricking the finger tip with the disposable pricking needle (Lancet, Monoject Scientific, Div. Sherwood Medical, U.K.). The filter paper was dried and kept for further analysis.

Chemical Analysis

The OCPs under investigation were α-, β-, γ- and δ-HCH (expressed as HCHs) and o,p'-DDE, p,p'-DDE, p,p'-DDD, o,p'-DDT and p,p'-DDT (expressed here as DDTs). The dried blood spots on the Whatman no.1 filter paper were cut into small pieces and immersed into 5ml of n-hexane: acetone mixture (1:1 v/v) for 20 minutes in a conical stoperred test tube and mixed thoroughly for 10 minutes with vortex mixer. Solvent phase was separated and the filter paper was re-extracted twice with 5 ml n-hexane: acetone mixture. The extracted solvent was pooled and concentrated to 2ml in vortex evaporator. Extraction recoveries of the HCH and DDT from spiked samples are more than 85 per cent which are tabulated in Table 7.1.

GC Analysis of Samples

Quantitative analysis of the OC-DDTs and OC-HCHs were undertaken on a Gas Chromatograph (GC) with Electron Capture Detector (Ni^{63}) (Perkin Elmer Clarus-500). The column used for the analysis was 25x.25 mm id. The oven temperature was programmed as an initial temperature of 150°C was held for 1 min and then the temperature was increased to 200°C at a rate of 5°C/min held for 5 min and finally to 275°C at a rate of 10°C/min. The injection port temperature was kept at 275°C and the Detector temperature was 300°C. Nitrogen was used as carrier gas at a constant flow of 1 ml/min. Nitrogen was also used as a makeup gas at 75 ml/min.

Table 7.1: Extraction Recoveries of HCH and DDT in Human Blood Samples

Insecticide	*% Recovery*	*Standard Deviation*	*Standard Error (S.E.)*	*% Coefficient of Variation*
α-HCH	90.9	0.6	0.3	0.6
β-HCH	90.5	0.5	0.2	0.5
γ-HCH	86.5	0.7	0.3	0.8
δ-HCH	92.4	0.3	0.1	0.3
o,p'-DDE	89.3	0.8	0.4	0.9
p,p'-DDE	87.6	0.8	0.3	0.9
p,p'-DDD	86.4	0.3	0.1	0.3
o,p'-DDT	85.1	0.7	0.3	0.8
p,p'-DDT	90.2	0.6	0.3	0.7

Statistical Analysis

All statistical analysis is carried out using the SPSS Microsoft version16.0 for Windows. Concentrations of OCPs in human blood are summarized using arithmetic means, standard deviations together with minimum and maximum values. Statistical analysis was conducted for all residues of DDT metabolites and HCH isomers in all investigated matrices.

Results and Discussion

Human blood is the most accessible body fluid for ascertaining the organochlorine residue levels. All the human blood samples were found to be contaminated with different levels of DDT and HCH residues. The mean concentration along with range values and standard deviation of DDT and HCH residues in analyzed blood samples are presented in Tables 7.2–7.5.

The OCPs measured in higher concentrations were p,p'-DDT (predominant contaminant of DDT), β-HCH (most persistent isomer of HCH) and α-HCH. In case of HCH isomers, it was observed that mean total concentrations were higher for district Nagaon, 627μg/

Table 7.2: Mean Concentration of HCH and their Isomers in the Human Blood Samples of District Dibrugarh

Sampling Sites	*No. of Samples*	*Residues in μg/L*				
		α-HCH X± SD Min-Max	*β-HCH X± SD Min-Max*	*γ-HCH X± SD Min-Max*	*δ-HCH X± SD Min-Max*	*Total HCHX X± SD Min-Max*
Tengakhat (D1)	31	177.40±183.1 (ND-765.80)	108.30±234.1 (ND-916.42)	55.70±64.6 (ND-259.46)	37.93±39.8 (ND-170.41)	379.93±308.4 (45.82-1208.18)
Dibrugarh (D2)	35	102.84±409.7 (ND-2430.62)	119.71±57.1 (ND-266.13)	44.64±58.9 (ND-216.39)	23.30±30.8 (ND-177.38)	297.68±411.9 (90.35-2541.88)
Joypur (D3)	33	130.37±84.1 (ND-312.14)	112.61±103.8 (3.35-388.02)	70.55±60.3 (2.82-236.51)	23.14±76.9 (ND-419.13)	336.68±186.0 (80.58-794.64)
Khowang (D4)	34	103.69±127.2 (4.42-688.57)	114.32±109.2 (41.53-532.00)	44.39±42.9 (ND-179.53)	43.57±27.1 (4.12-124.33)	305.97±173.6 (112.02-822.50)
Balijan (D5)	36	235.33±87.4 (ND-504.33)	137.86±300.4 (ND-1616.41)	32.90±37.3 (0.53-148.01)	14.45±24.8 (ND-123.60)	420.55±359.8 (119.88-1995.96)
Mean HCH	**169**	**150.29±220.8**	**119.01±183.7**	**49.18±54.4**	**28.14±44.6**	**348.11±305.0**

Table 7.3: Mean Concentration of DDT and their Metabolites in the Human Blood Samples of District Dibrugarh

Sampling Sites	*No. of Samples*	*Residues in µg/L*					
		o,p-DDE X± SD Min-Max	*p,p-DDE X± SD Min-Max*	*p,p-DDD X± SD Min-Max*	*o,p-DDT X± SD Min-Max*	*p,pDDT X± SD Min-Max*	*Total DDT X± SD Min-Max*
Tengakhat (D1)	31	16.49±43.5 (ND-206.38)	96.12±149.0 (ND-688.46)	49.38±116.1 (ND-595.40)	56.89±97.3 (ND-350.71)	147.54±214.0 (ND-1054.54)	366.11±400.3 (16.64-1836.20)
Dibrugarh (D2)	35	18.90±26.8 (ND-91.00)	40.24±73.2 (ND-354.12)	58.92±139.3 (ND-734.00)	45.71±69.1 (ND-259.66)	153.38±158.7 (1.37-686.68)	313.94±268.3 (3.57-1094.29)
Joypur (D3)	33	15.81±21.9 (ND-71.20)	38.10±45.5 (ND-165.37)	38.61±53.8 (ND-266.09)	36.75±83.7 (ND-417.09)	219.58±814.8 (ND-4634.20)	348.85±820.3 (10.98-4718.16)
Khowang (D4)	34	23.49±29.7 (ND-95.30)	28.44±34.6 (ND-161.27)	40.03±100.6 (ND-414.06)	15.19±18.6 (ND-76.98)	235.44±129.3 (ND-513.83)	342.59±148.6 (94.04-702.07)
Balijan (D5)	36	13.28±24.6 (ND-96.90)	132.65±248.1 (ND-1133.28)	53.00±108.9 (0.34-530.94)	61.35±112.8 (ND-507.06)	432.85±294.9 (ND-1288.81)	693.12±472.7 (14.73-2063.80)
Mean DDT	**169**	**17.58±29.8**	**67.38±142.1**	**47.79±106.8**	**43.20±83.5**	**241.27±415.1**	**416.86±493.3**

Table 7.4: Mean Concentration of HCH and their Isomers in the Human Blood Samples of District Nagaon

Sampling Sites	*No. of Samples*	*Residues in μg/L*				
		α-HCH X± SD Min-Max	*β-HCH X± SD Min-Max*	*γ-HCH X± SD Min-Max*	*δ-HCH X± SD Min-Max*	*Total HCHX X± SD Min-Max*
Dhing (N1)	31	264.82±206.4 (ND-650.89)	212.82±174.6 (ND-561.30)	140.91±261.5 (ND-1442.53)	100.53±147.3 (ND-661.70)	719.08±461.3 (60.76-2432.55)
Nagaon (N2)	35	110.68±308.3 (ND-1767.49)	177.08±263.9 (ND-1093.91)	130.17±319.8 (ND-1831.64)	57.75±71.1 (ND-202.21)	475.68±724.9 (62.47-3278.52)
Koliabor (N3)	33	200.50±257.2 (ND-1164.76)	176.30±147.8 (ND-519.97)	126.20±278.1 (ND-1589.55)	59.03±118.9 (ND-459.86)	562.02±581.3 (66.97-3214.17)
Hojai (N4)	32	180.99±143.5 (35.35-513.96)	253.69±124.4 (ND-525.03)	110.15±210.3 (ND-1064.22)	66.35±179.7 (ND-1005.98)	611.17±488.3 (199.73-2810.48)
Lumding (N5)	31	305.58±402.0 (0.02-2169.69)	274.38±181.0 (ND-727.57)	157.59±183.2 (0.42-707.70)	48.68±67.3 (ND-272.06)	786.23±503.6 (198.57-2921.75)
Mean HCH	**162**	**210.76±281.8**	**217.50±186.6**	**132.66±253.8**	**66.18±124.1**	**627.10±566.4**

Table 7.5: Mean Concentration of DDT and their Metabolites in the Human Blood Samples of District Nagaon

Sampling Sites	*No. of Samples*	*Residues in µg/L*					
		o,p-DDE X± SD Min-Max	*p,p-DDE X± SD Min-Max*	*p,p-DDD X± SD Min-Max*	*o,p-DDT X± SD Min-Max*	*p,pDDT X± SD Min-Max*	*Total DDT X± SD Min-Max*
Dhing (N1)	31	58.61±61.2 (ND-196.24)	144.74±371.7 (ND-2083.29)	55.87±69.5 (ND-304.33)	191.28±359.2 (ND-1504.75)	326.38±398.0 (ND-1633.67)	786.87±792.0 (45.61-3509.19)
Nagaon (N2)	35	15.93±56.7 (ND-320.94)	123.24±99.9 (2.73-367.49)	28.87±47.1 (ND-223.72)	206.67±505.6 (0.79-2844.41)	241.65±568.4 (ND-2863.74)	616.36±1011.7 (58.39-4917.83)
Koliabor (N3)	33	19.38±67.1 (ND-385.52)	179.32±640.5 (ND-3700.03)	82.82±123.9 (ND-604.22)	246.31±829.1 (ND-4769.56)	371.89±435.1 (ND-2097.70)	899.73±1728.4 (6.94-9906.42)
Hojai (N4)	32	93.51±127.4 (ND-578.56)	99.12±135.4 (ND-659.28)	176.93±326.8 (ND-1528.38)	48.57±84.3 (ND-411.47)	239.68±484.7 (ND-2768.65)	666.92±571.4 (75.09-2882.74)
Lumding (N5)	31	37.60±56.2 (ND-297.35)	179.62±434.0 (ND-1933.46)	76.54±117.7 (ND-610.35)	119.78±195.6 (ND-1025.97)	343.36±303.1 (ND-1138.47)	756.89±616.9 (24.04-3079.87)
Mean DDT	**162**	**45.71±84.2**	**144.35±386.3**	**86.14±179.3**	**161.01±472.9**	**303.43±446.2**	**742.61±1028.3**

L as compared to district Dibrugarh, 348.11µg/L. In Nagaon population, β-HCH is the prevalent isomer of HCH, with a mean value of 217.5 µg/L while in population of Dibrugarh, it was α isomer that contributed maximum with a mean value of 150.29 µg/L. γ isomer is also present in significant amounts with mean concentrations of 132.66 µg/L and 49.18 µg/L for Nagaon and Dibrugarh respectively. Presence of significant amounts of α-HCH isomer and high levels of β-HCH isomer found in all the samples suggests that the general population is exposed to technical HCH. In addition, γ-HCH with high values and high detection frequency (90.12 per cent and 94.08 per cent samples for Nagaon and Dibrugarh) is indicating simultaneous use of pure lindane pesticide in both the districts.

All the blood samples showed the presence of some type of DDT residues, the most frequently detected was p,p'-DDT which was present in (90.74 per cent) and (88.76 per cent) of samples followed by o,p'-DDT (82.72 per cent) for Nagaon and p,p'-DDE (84.62 per cent) for Dibrugarh respectively. Among DDT metabolites, higher concentrations were observed in District Nagaon, 742.61 µg/L than Dibrugarh, 416.86µg/L. It has been noticed that the major contributor to Σ DDT is p,p'-DDT with mean concentration of 303.43 µg/L and 241.27 µg/L and alone contributed more than 40 per cent and 57 per cent of total DDT contamination in Nagaon and Dibrugarh population respectively.

The next major contaminant was o,p'-DDT, 161.01 µg/L (21 per cent) in Nagaon district while in Dibrugarh district, it is p,p à dEb-DDE with a mean concentration of 67.38 µg/L (16 per cent). o,p'-DDE metabolite was detected at lower levels with a mean value of 45.71 µg/L and 17.58 µg/L.

From analyzing the statistical data on mean concentrations, most significant differences have been found for β-HCH, for γ-HCH, for total HCH, for p,p'-DDD, for o,p'-DDT and for total DDT ($p=0.000$) while less significant for p,p'-DDE ($p=0.019$) between both the districts.

The present study is reporting levels of DDTs and HCHs residues in representative samples of population residing in two districts of Assam state, North-east region of India for the first time. OCPs have contributed greatly to the increase of yields in agriculture by controlling the pests and diseases and also towards checking the insect borne diseases (malaria, dengue, filariasis and encephalitis etc.) in the public health sector (Bhatnagar, 2001; Rekha *et al.*, 2006). Through misuse and overuse, there is considerable pollution contributing to the adverse environmental and health consequences. Inappropriate application of pesticides is affecting the whole ecosystem by entering the residues in human system through food chain and polluting the soil, surface and groundwater and air (Agnihotri, 1999; UN/DESA, 2002).

The results show that p,p′-DDT, the parental compound of technical DDT turned out to be the most frequently detected residue in both the districts (91 per cent blood samples for Nagaon and 89 per cent for Dibrugarh). p,p′-DDE comes next with 85 per cent frequency in Dibrugarh while in Nagaon, it was o,p′-DDT in about 83 per cent samples. It is important to notice that human population from Assam shows high levels of p,p′-DDT and o,p′-DDT, the more estrogenic forms of DDT. p,p′-DDE also present in significant amounts indicates breaking down of p,p′-DDT at a relatively rapid rate in the environment to become the persistent form of this pesticide (Ecobichon, 1995; Safe, 1994). The levels of p,p′-DDE are used frequently for evaluation of chronic exposure to DDT. Although p,p′-DDE is often used as substitute of past exposure to all the sources of DDT and DDE, there is no clear way to investigate that how much amount of p,p′-DDE present in human blood is a result of direct exposure to sprayed technical DDT and hence to the more estrogenic forms of DDT - p,p′-DDT and o,p′-DDT or due to ingestion of dietary performed p,p′-DDE.

In an order to differentiate between current and historical exposure to DDT, DDT/DDE ratio has been used frequently (Ahlborg *et al.*, 1995; Jaga and Dharmani, 2003). A low DDT/DDE ratio indicates high persistence in the environment and ongoing biomagnifications (Jaga and Dharmani, 2003) whereas a high DDT/DDE ratio implies both unceasing and also ongoing exposure to DDT. In the present study, DDT/DDE ratio is found to be very high up to 3.6 for district Dibrugarh and 2.1 for district Nagaon, thus indicating that there is an active and continuous source of DDT in the area in large amounts. These values are extremely high and only have been reported previously from region/countries that are using DDT currently for vector control in public health and/or agriculture, such as Mexico and Tanzania (these are the countries where population are subjected to high technical DDT exposure). But the results reported about these countries were observed for fat tissue not for the blood samples (Rivero-Rodriguez *et al.*, 1997; Vander Ven *et al.*, 1992). These results have been compared together as both the fat tissue and blood are the part of human system.

The results have shown that γ-HCH is the frequently detected (90 per cent blood samples for Nagaon and 94 per cent for Dibrugarh) among four HCH isomers. Mean total HCH value is higher in district Nagaon with maximum contribution of β isomer followed by α isomer. α and β isomers are essentially non-insecticidal, they appear probably as a result of impure lindane formulation or application of technical HCH in agriculture, although technical HCH has been banned in agriculture in India since 1997. A high level of these two isomers is of concern, especially β-HCH, due to its high persistence in environment as well as in human system (Ahlborg *et al.*, 1995; Lino and Silviera, 1996). Presence of γ-HCH concentration in considerable amounts indicates the application of lindane in agriculture. Among HCH isomers, high levels of α and γ isomers indicates the recent exposure while high β isomer values represent old exposure due to its high persistence (Suntio *et al.*, 1988). These results are consistent with the fact that cumulative consumption of HCHs in India until 1985 was 575,000 tons and since then about 45,000 tons of HCHs has been used annually (Voldener and Li, 1995).

In India, areas under monoculture and continuous cultivation of paddy, cotton and other commercial crops are increasing. Also, pesticide use is high in regions with good irrigation facilities. Paddy is the second largest crop after cotton which is grown on 24 per cent of the cultivated area and consumes about 25 per cent of the total pesticides used in the country (Abhilash and Singh, 2009). In Dibrugarh and Nagaon, a major part around 81553 hectare (24.09 per cent) and 160035 hectare (38.94 per cent) area is under paddy cultivation with 166309 hectare (49.12 per cent) and 354801 hectare (86.32 per cent) total sown area (NIC, 2004-05) which is subjected to high application of HCH pesticide may be a possible explanation for high HCH concentration in those districts.

Conclusions

Population from both the districts contained much higher levels of these contaminants in human blood samples. Such higher levels may have resulted from ongoing exposure to these chemical substances which are being used in this region. These findings should be a matter of public health concern as the OCPs residues go beyond the occupational exposure and affects the general population living in areas subjected to intensive application of DDT and HCH for vector control in public health and for agricultural activities. As health consequences of these high residue levels in human blood remain uncertain, future monitoring studies are required in an attempt to assess the relation between contamination levels and their undesired effects on human health and to environment. There is an urgent need to apply less persistent and biodegradable pesticides in this region to reduce the environmental pollution and thus the risk due to their accumulation in human body.

References

Abhilash, P.C. and Singh, N. 2009. Pesticide use and application: An Indian Scenario. *J. Hazard. Mater.* 165:1-12.

Agnihotri, N.P. 1999. *Pesticide Safety and Monitoring, All India Coordinated Research Project on Pesticide Residues.* Indian Council of Agricultural Research, New Delhi, India.

Ahmed, M.T., Loutfy, N. and El Shieke, E. 2002. Residue levels of DDE and PCBs in blood serum of women in Port Said in Egypt. *J. Hazard. Mater.* 89:41-48.

Amaral Mendes, J.J. 2002. The endocrine disruptors: a major medical challenge. *Food Chem. Toxicol.* 40:781-788.

Bhatnagar, V.K. 2001. Pesticides pollution: trends and perspectives. *ICMR Bull.* 31:87-88.

Brody, J.G. and Rudel, R.A. 2003. Environmental pollutants and breast cancer. Environ. *Health Perspect.* 111:1007-1119.

Calle, E.E., Frumkin, H., Henley S.J., Savitz, D.A., Thun, M.J. 2002. Organochlorines and breast cancer risk. *C. A. Cancer J. Clin.* 52:301-309.

Charlier, C.J., Plomteux, G.J., 2002. Determination of organochlorine pesticides residues in the blood of healthy individuals. Clin. Chem. Lab. Med.40 (4):361-364.

Ceron, J.J., Gutierrez-Panizo, C., Barba, A., Camara, M.A. 1995. Endosulfan isomers and metabolite residue degradation in carnation (*Dianthus caryophyllus*) by product under different environmental conditions. *J. Environ. Sci. Health.* 30:221-232.

Dua, V.K., Pant, C.S., Sharma, V.P., Pathak, G.K. 1996. Determination of HCH and DDT in finger-prick whole blood dried on filter paper and its field application for monitoring concentration in blood. Bull. Environ. *Contam. Toxicol.* 56:50-57.

Edwards, T.M., Moore B.C., Guillette, L.J. Jr. 2006. Reproductive dysgenesis in wildlife: a comparative view. *Int. J. Androl.* 29(1):109–121.

Frias, M.M., Frenich, A.G., Vidal, J.L.M., Sanchez, M.M., Olea, F., Olea, N. 2001. Analysis of lindane, vinclozolin, aldrin, p,p′-DDE, o,p′-DDT and p,p′-DDT in human serum using gas chrimatography with electron capture detection and tandem mass spectrometry. *J. Chromatogr.* B 760:1-15.

Gammon, M.D., Wolff, M.S., Neugut, A.I., Eng, S.M., Teitelbaum, S.L., Britton, J.A., Terry, M.B., Levin, B., Stellman, S.D., Kabat, G.C., Hatch, M., Senie, R., Berkowitz, G., Bradlow, H.L., Garbowski, G., Maffeo, C., Montalvan, P., Kemeny, M., Citron, M., Schnabel, F., Schuss, A., Hajdu, S., Vinceguerra, V., Niguidula, N., Ireland, K., Santella, R.M. 2002. Environmental toxins and breast cancer on Long Island. II. Organochlorine compound levels in blood. *Cancer Epidemiol. Biomarker. Prev.* 11, 686-697.

Gupta, P.K. 2004. Pesticide exposure-Indian Scene. *Toxicol.* 198:83-90.

Hansen, J.C. 1998. The human health programme under AMAP. AMAP human health group. Arctic monitoring and assessment program. *Int. J. Circumpolar Health.* 57:280-291.

Herrman, J.L. 1993. The role of the World Health Organization in the evaluation of pesticides. Regul. Toxicol. Pharmacol. 17:282-286.

Jaga, K., Dharmani, Ch. 2003. Global Surveillance of DDT and DDE levels in Human tissues. *Int. J. Occup. Environ. Health.* 16:7-20.

John, P.J., Bakore, N., Bhatnagar, P. 2001. Assessment of organochlorine pesticide residue levels in dairy milk and buffalo milk from Jaipur City, Rajasthan, India. *Environ. Int.* 26:231-236.

Kunisue, T., Someya, M., Kayama, F., Jin, Y., Tanabe, S. 2004. Persistent organochlorines in human breast milk collected from primiparae in Dalian and Shenyang. *China Environ. Pollut.* 131:381-392.

Lallas, P. 2001. The Stockholm Convention on persistent organic pollutants. *Am. J. Int. Law.* 95:692-708.

Lino, C.M., Silveira, M.I.N. 1996. Contribuicoes para a presenca de residuos de pesticidas organoclorados no Homem. *Rev. Port. Farm.* 46 (4):143-147.

Mukherjee, I. and Gopal, M. 2003. Organochlorine insecticide residues in drinking water and ground water in and around Delhi. *Environ. Monit. Assess.* **76:**185–193.

Rekha, S.N., Naik, R., Prasad. 2006. Pesticide residue in organic and conventional food-risk anlysis. *Chem. Health Safety.* 13:12-19.

Rivero-Rodriguez, L., Borja-Aburto, V.H., Santos-Burgoa, C., Waliszewskiy, S., Rios, C., Cruz, V. 1997. Exposure Assessment for workers applying DDT to control malaria in Veracruz, Mexico. *Environ. Health Perspect.* 105:98-101.

Snedeker, S. 2001. Pesticides and breast cancer risk: a review of DDT, DDE and dieldrin. *Environ. Health Perspect.* 109 (1):35-47.

Suntia, L.R., Shiu, W.Y., Mackay, D., Seiber, J.N., Glotfelty, D. 1988. Critical review of Henry's law constants for pesticides. *Rev. Environ. Contam. Toxicol.* 103:1-59.

UN/DESA. 2002. Changing Unsustainable Patterns of Consumption and Production, Johannesburg Plan on Implementation of the World Summit on Sustainable Development, Johannesburg (Chapter III).

Vander Ven, K., Vander Ven, H., Thibold, A., Bauer, O., Kaisi, M., Mbura, J. 1992. Chlorinated hydrocarbon content of fetal and maternal body tissues and fluids in full term pregnant women. A comparison of Germany vs. Tanzania. *Hum. Reprod.* 7:95-100.

Wania, F. andMackay, D. 1996. Tracking the distribution of persistent organic pollutants. *Environ. Sci. Technol.* 30:390A–396A.

Wolff, M.S. Zeleniuch-Jacquotte, A., Dubin, N., Toniolo, P. 2000. Risk of breast cancer and organochlorine exposure. Cancer Epidemiol. *Biomarker Prev.* 9:271-277.

Chapter 8

Oil Spill in Ocean: Control and Management

Kanchan Deoli Bahukhandi, S.C. Gupta, V. Sujatha and Pawan Mandappa

Oil spill is one of the major causes of marine pollution which is directly responsible for depletion of aquatic biodiversity. The term often refers to marine oil spills, where oil is released into the ocean or coastal waters. The oil may be a variety of materials including crude oil, refined petroleum products (such as gasoline or diesel fuel) or by-products, ships' bunkers, oil refuse or oil mix in waste. There are several instances of oil spill in the past which severely deteriorated a marine ecosystem. There is need to control marine pollution caused by oil spill. In this Chapter an efforts has been made to highlight the impact of disastrous oil spill which occurred in the past and proposes various methods for controlling oil spill.

Introduction

Today the whole world is facing problem of environmental crisis. These are natural, *i.e.* earthquake, landslide, cyclone, Catherina, thunder, lightning as well as man-made environmental problem, *i.e.* environmental pollution, global warming, ozone layer depletion, acid rain etc. Instead of developing good disaster management plan and restoration of environmentally fragile area, we can't put much effort to control natural environmental problem, but the man-made environmental problem can be controlled by changing our life styles and implementing sound engineering control practices. When any industry is established there are always possibilities of four type of water pollution air, water, soil and noise pollution which affect whole biodiversity and degrade natural resources. Oil spill is one of the major causes of water pollution and it changes physical, chemical and biological characteristics of water which become unfit for consumption of marine life. The term often refers to marine oil spills, where oil is released into the ocean or coastal waters. The oil may be a variety of materials including crude oil, refined petroleum products (such as gasoline or diesel fuel) or by-products, ships' bunkers, oil refuse or oil mix in waste.

Oil spills happen when people make mistakes or are careless and cause an oil tanker to leak oil into the ocean. There are a few more ways an oil spill can occur, *i.e.* equipment breaking down and when countries are at war, one country may decide to dump gallons of oil into the other country's oceans; for example the incident of Gulf war in which the massive oil slick in the Gulf happened in February 1990 due to Iran, Iraq war. Oil spill hit the Saudi Shores and is and threatened the shore of Bahrain, Qatar, Kuwait and United Urab Emigrates. It has been estimated that millions of birds were killed and apart from killing birds and marine life, the oil also threatens to disrupt the desalination plants on the western coast of Gulf that provide 50 per cent water to Arabian states. Terrorists may cause an oil spill because they dump oil into a country's ocean. Illegal dumpers are people who dump crude oil into the oceans. Natural disasters (like hurricanes) may cause an oil spill, too. If a hurricane was a couple of miles away, the winds from the hurricane could cause the oil tanker to flip over, pouring oil out.

Major Oil Spills of the World

In the past there were lots of incidents of oil spills which severely degraded aquatic ecosystem; for example the Santa Barbara oil spill occurred in 1969 in the Santa Barbara Channel, which is a part of the Pacific Ocean off the coast of Southern California, United States of America due blow-out on Union Oils Platform A, six miles offshore, in the Dos Cuadras Offshore Oil Field. Over a ten-day period, an estimated 80,000 to 100,000 barrels of crude oil spilled into the channel and onto the beaches of Santa Barbara County in Southern California, fouling the coastline from Goleta to the Rincon and all four of the northern Channel Islands. It has been estimated around 10,000 birds were killed in the ecological disaster. The incident of Bombay High Oil Slick damaged the Ecosystem. On May 17, 1993, the Uran oil pipeline ruptured at an offshore platform causing a 4 km long and 400 meters wide massive oil spill in Bombay which severely affects flora and fauna of marine environment.

Torry Canyon Accident the world first disastrous consequences of an oil spill occurred on 18 March 1967 near the entrance of English channel, spilling 60,000 tonnes of crude oil into the sea. Countless number of fishes, birds and aquatic fauna were killed. The Exxon Valdez incident was a major oil spill. This oil spill took place on March 23, 1989 at Prince William Sound in Alaska. The drivers of the Exxon Valdez noticed icebergs in their planned route. Instead of trying to weave through the icebergs, they decided to go in another direction. On the new route, the oil tanker hit shallow land. The shallow land was not a beach, it was called Bligh Reef and incident had taken place. The spill greatly affected sea otters and sea birds. There were about 2,800 sea otters and 250,000 seabirds killed by the spill. Harbor seals, bald eagles, killer whales, and salmon were also greatly affected by the spill. The diagram shows where the oil flowed and how far it flowed through Alaska. The map also tells you where the oil spill started in Alaska, which would be in Prince William Sound.

There has been an oil spill on May 9, 2001 in the Galapagos Islands this Islands are located off the coast of Ecuador. The tanker, Jessica, tipped over pouring 240,000 gallons of oil into the ocean. Chemicals had been used to break down the oil. It has been observed that if the oil is able to reach the bottom of the ocean, it might take at least two years to clean. Sea lions, pelican, seagulls, sea urchins, and seaweed have been affected by this spill. Soap is used to clean the fur and feathers of hurt animals and milk for the animals' heads because it wouldn't hurt animals eyes.

The Sea Star Oil Spill 1972 the South Korean supertanker, Sea Star, collided with a Brazilian tanker, the Horta Barbosa, off the coast of Oman on the morning of Dec. 19, 1972. The vessels caught fire after the collision and the crew abandoned ship. Although the Horta Barbosa was extinguished in a day, the Sea Star sank into the Gulf on Dec. 24 following several explosions.

Figure 8.1

Day 7: 90 miles; Day 11: 140 miles; Day 15: 180 miles; Day 19: 250 miles; Day 38: 280 miles; Day 40: 350 miles; Day 56: 470 miles

GREEN=The oil; STAR = Prince William Sound

Odyssey Oil Spill 1988, this large oil spill occurred about 700 nautical miles off the coast of Newfoundland and spilled more than 40 million gallons of oil into the ocean. Haven Tanker Oil Spill has taken place in the year 1991 and about 45 million gallons were spilled due explosion of Oil Tanker and sank off the coast of Italy. It has been estimated that six people were killed and leaking its remaining oil into the Mediterranean for 12 years. ABT Summer Oil Spill spilled 51-81 million gallons due to explosion of ship of the coast of Angola, discharging massive amounts of oil into the ocean in the year 1991.

Amoco Cadiz Oil Spill occurred in the year March 16, 1978 in Portsall, France spilling about 69 million gallons oil. The massive Amoco Cadiz was caught in a winter storm that damaged the ship's rudder. Castillo de Bellver Oil Spill had taken place in the year 1983 at Saldanha Bay, South Africa and spilled 79 million gallons oil. Nowruz Oil Field Spill, 1983 was the result of a tanker collision with an oil platform. The weakened platform was closed, and it collapsed upon impact, spewing oil into the Persian Gulf. The Kolva River Oil Spill 1983 caused due to splilling of 84 million gallons oil in Kolva River, Russia. Atlantic Empress Oil Spill in: July 19, 1979 spilling 90 million gallons at off the coast of Trinidad and Tobago. Ixtoc 1 Oil Spill, 1979 occurred at Bay of Campeche off Ciudad del Carmen, Mexico and spilled 140 million gallons oil.

Control and Management of Oil Spill

There are many ways to stop the spread of oil in the ocean. Workers can place a boom around the tanker that is spilling oil. Booms collect the oil off the water. Skimmers are used to remove oil. Chemicals are also used into ocean which can breakdown oil into ocean. Following are the techniques used for controlling oil spill into the ocean.

Boom

Some of the tools used to control oil in a spill include 'booms', which are floating barriers used to clean oil from the surface of water. A boom can be placed around the tanker that is spilling oil. Booms collect the oil off the water. A boom may be placed somewhere before an oil spill. They can also be placed around an entrance to the ocean, like a stream. They also can be placed around a habitat with many animals living there. These booms absorb any oil that flows around it.

Skimmer

They are also used as skimmers to control oil spill. Skimmers are boats that can remove the oil off the water. Skimmers use pumps or vacuums to remove oil as it float on water.

Sorbents

Sorbent booms and barriers are used to absorb a moving oil slick. They only work well when a slick is thin, because once their surfaces are saturated, they can't absorb anymore. Sorbents recover oil either through absorption (the oil is drawn into porous materials) or adsorption (the oil sticks to the surface of the sorbents). They're used in the final mopping up because they get at trace amounts of oil left after skimming and can be sent into areas where skimmers can't reach. Sorbents come in two basic types: natural organic materials like peat moss and sawdust; and synthetic organic sorbents like polypropylene, polyester foam, polystyrene and polyurethane. Both types are usually applied by hand, and recovered with nets, rakes, forks and pike poles.

Dispersant

Dispersants are chemicals, which have components of surface-active agents called surfactants. The dispersants aids in the breaking up of the oil slick into smaller droplets. Chemical dispersants remove the oil from the surface of the water and into the water column by enhancing the natural chemical and physical breakdown of oil. Once in the water column, the dilution of oil is greatly facilitated and hence the toxicity level is reduced. Eventually, these dispersants and oil droplets are food sources for bacteria in the ocean. By removing the oil from the water surface, birds, marine mammals, turtles, and sensitive coasts are protected. Chemical dispersants do not cause the oil to sink, but remain in suspension in the water column. The disadvantages of chemical dispersants are that they are toxic in certain ways. As an added precaution, chemical dispersants are not applied to shallow near-shore waters, mangrove areas, marshes, or waters over coral reefs and sea grass beds.

Bioremediation

Oil, like many natural substances, biodegrades over a period of time into simple compounds such as carbon dioxide, water and biomass. Bioremediation is the term used to describe a range of processes, which can be used to accelerate natural biodegradation. Biodegradation occurs as a result of the oxidation of certain components of spilled oil by microbes such as bacteria, fungi, unicellular algae and protozoa. The rate at which this natural process occurs is limited by several factors including the temperature, and the levels of microbes, nutrients and oxygen present in the immediate environment. Other factors such as the chemical composition and the amount of weathering of the spilled oil are also important. Although Bioremediation presents a good picture it is a very slow process and takes a long time to work fully. It can work effectively only on oil which is spread as a thin layer on water. Bioremediation is not suitable for removing large amounts of oil and should only be considered where the concentration of oil is low as a final polishing technique.

In-situ Burning

Occasionally the slick caused by a spill is removed through a controlled burn. Workers can burn freshly spilled oil with fireproof booms to contain the oil. Burning, only works under certain wind and weather conditions. Burning is the last option to decide, as this method causes air pollution.

Oil Spill Control Manually

Oil spill control manually on land is often conducted manually. Scooping, cleansing and scraping of the rocks and sand is performed until the oil has been removed. People can use high or low pressure hoses to spray the oil that is on the beaches. Vacuum trucks may be driven on the beaches to vacuum up the oil. They can also simply use shovels or road equipment to collect all the oil off the beaches.

Biotechnology in Treating Petroleum Sludge and Oil Spills

Biotechnology, however, has shown that particular species of bacteria and fungi, normally found in soil, can protect the marine environment by breaking down various types of hydrocarbons, the main component of petroleum. These bacteria are structure specific, making it hard for one microorganism to clean up a mix of oil. A genetically modified organism that can break down multiple compounds could be used to clean spills very efficiently. A bacterium like *Alcanivorax borkumensis* could provide the base for such an organism.

Proper Disposal of Used Oil

Used oil should be put in a clean plastic container with a tight lid. It should not be mixed with any thing else – paint gasoline, solvents, antifreeze, etc. that will make it unsuitable for recycling. Used oil should be taken for recycling center that collects used oil, a service station, quick lube, or any location that accepts used oil.

Illegal Disposal Practices

- Pouring used oil down a drain.
- Pouring used oil into a storm sewer.
- Tossing used oil on your street or the ground.
- Disposing of oil in lakes, streams or wetlands.

Different Types of Sensors for Oil Spill Monitoring

Optical techniques are the most common means of remote sensing. Cameras, both still and video are common because of their low price. In recent years, visual or camera observation has been enhanced by the use of GPS (Global Positioning Systems). Direct annotation of video images with GPS information is possible and provides useful documentation.The two basic types of radar that can be used to detect oil spills and for environmental remote sensing in general are Synthetic Aperture Radar (SAR) and Side-Looking Airborne Radar (SLAR).

Recommendations

- ✰ Well-planned safety inspections (within every 3 months) help in detecting spill oil before an accident occurs.
- ✰ Before an inspection, analyze past accidents to determine specific causes and high hazard areas or operations.
- ✰ Strict rules, regulation, and law should be followed.
- ✰ Safety awareness of employee.
- ✰ Equipments should be well maintained.

References

Amber. "Sea Otter Rehabilitation." http://www.kpbsd.k12.ak.us/seward.elem/ocean/seaotter.html Last Visited: January, 2002.

Barid, Stuart. "Oil Spills." http://www.iclei.org/efacts/oilspi2.gif Last Visited: January, 2002.

http://www.appsi.com

http://www.gri.org

Office of Response and Restoration, Revised: November 30, 2000. Office of Response and Restoration, National Ocean Service, National Oceanic and Atmospheric Administration

Oil Spills: An Initial Response Manual for Attorneys General - Douglas K. Mertzff http://www.alaska.net/~dkmertz

Roy, W. Hann Jr. *Oil Spill Prevention and Control for the 21st Century.* Center for Oil Spill Technology, Texas Engineering Experiment Station, 1994.

Roy, W. Hann Jr. Oil Spill Prevention and Control for the 90's. Center for Oil Spill Technology, Texas Engineering Experiment Station, 1994.

TAMU Spill Incident Management System (SIMS). Center for Oil Spill Technology, Texas Engineering Experiment Station, Dr. Roy W. Hann Jr. and George Alan Frantz

The Charles Darwin Foundation. "Technical Chronology of CDRS Actions." http://www.darwinfoundation.org/oilspill/oilspillmay9.html Last Visited: January, 2002.

Part-III

CLIMATE CHANGE AND WATER RESOURCES

Chapter 9

Impacts of Climate Change on Water

Bhanu Chaddha and Vaibhav Matta

This Chapter describes the multidimensional impact of climate change on global water resources. Observational records and climate projections provide abundant evidence that freshwater resource are vulnerable and have the potential to be strongly impacted by climate change. Observed warming over several decades has been linked to changes in the large-scale hydrological cycle such as: increasing atmospheric water vapor content; changing precipitation patterns, intensity and extremes; reduced snow cover and widespread melting of ice; and changes in soil moisture and runoff. There have been significant decreases in water storage in mountain glaciers and Northern Hemisphere snow cover. Shifts in the amplitude and timing of runoff in glacier- and snowmelt-fed rivers, and in ice-related phenomena in rivers and lakes, have been observed. By the middle of the 21st century, annual average river runoff and water availability are projected to increase as a result of climate change at high latitudes and in some wet tropical areas, and decrease over some dry regions at mid-latitudes and in the dry tropics. Increased precipitation intensity and variability are projected to increase the risks of flooding and drought in many areas. Water supplies stored in glaciers and snow cover are projected to decline in the course of the century, thus reducing water availability during warm and dry periods in regions supplied by melt water from major mountain ranges, where more than one-sixth of the world's population currently live. Higher water temperatures and changes in extremes, including floods and droughts, are projected to affect water quality and exacerbate many forms of water pollution from sediments, nutrients, dissolved organic carbon, pathogens, pesticides and salt, as well as thermal pollution with possible negative impacts on ecosystems, human health, and water system reliability and operating costs. Changes in water quantity and quality due to climate change are expected to affect food availability, stability, access and utilization. This is expected to lead to decreased food security and increased vulnerability of poor rural farmers. Globally, water demand will grow in the coming decades, primarily due to population growth and increasing affluence; regionally, large changes in irrigation water demand as a result of climate change are expected. Mitigation measures can reduce the magnitude of impacts of global warming on water resources, in turn reducing adaptation needs. However, they can have considerable negative side effects, such as increased water requirements for afforestation/reforestation activities or bio-energy crops.

Introduction

Scientists around the world now agree that the climatic changes occurring internationally are the result of human activity. Although responsibility for the causes of climate change rests primarily with the developed and industrialized nations, In particular, changes to water quality, quantity and availability will be an impact of ongoing climate change in many areas.

What Is Climate Change?

The United Nations Framework Convention on Climate Change (UNFCCC) defines climate change as, "a change of climate which is attributed directly or indirectly to human activity that alters the composition of the global atmosphere and which is in addition to natural climate variability observed over comparable time periods". Climate change is caused by greenhouse gasses (GHGs), which enhance the "greenhouse" properties of the earth's atmosphere. These gasses allow solar radiation from the sun to travel through the atmosphere but prevent the reflected heat from escaping back into space. This causes the earth's temperature to rise. GHG emissions have been rising since industrialization in the 1900s, due to increased burning of fossil fuels. Further significant increases in GHG levels are expected, particularly as developing countries become more industrialized. Based on present trends, the IEA World Energy Outlook predicts a 53 percent increase in global primary energy demand by 2030, with 70 percent of that coming from developing countries. As emerging economies, such as China and India, grow their contribution to energy demand will account for an increasing proportion of the total. Fossil fuels are expected to dominate this increase, and the subsequent GHG emissions will in turn lead to rising temperatures.

The EU has defined dangerous climate change as an increase in 2 degrees Celsius of average global temperatures. Since 1900, global temperatures have risen by 0.7 degrees Celsius and are continuing to rise at an estimated rate of 0.2 degrees per decade. If left unchecked, this implies global warming of at least 1.4 degrees Celsius.

More recent scientific research finds that these temperature projections are actually much higher than previously thought. GHG emissions are not remaining stable – they are increasing due to increased energy consumption as a result of population growth and industrialization. From a pre-industrial concentration of 280 ppm, the current stock of greenhouse gases in the atmosphere is 430 ppm, and is increasing by approximately 2.5 ppm per year. Even if the world's GHG emissions stabilize at 550 ppm, it is estimated that, at a minimum, there is a 63 percent chance of exceeding the declared 'dangerous' limit of 2 degrees Celsius temperature increase.

The effects of climate change on temperature are demonstrated by the "Hockey Stick" graph (see Figure 9.1). The graph, which measures temperature variation over the last 1,000 years, indicates a sharp rise in temperatures post-1900. The graph has been used to illustrate the urgent need for action on reducing GHGs to policymakers.

Impact of Climate Change

The effects of minor levels of climate change are already being felt, with impacts across many economic sectors. While there will clearly be some gains from climate change (for example, agriculture in some northern regions should increase in productivity due to a rise in temperatures), most of the impacts will be negative, and gains and losses will not be evenly distributed.

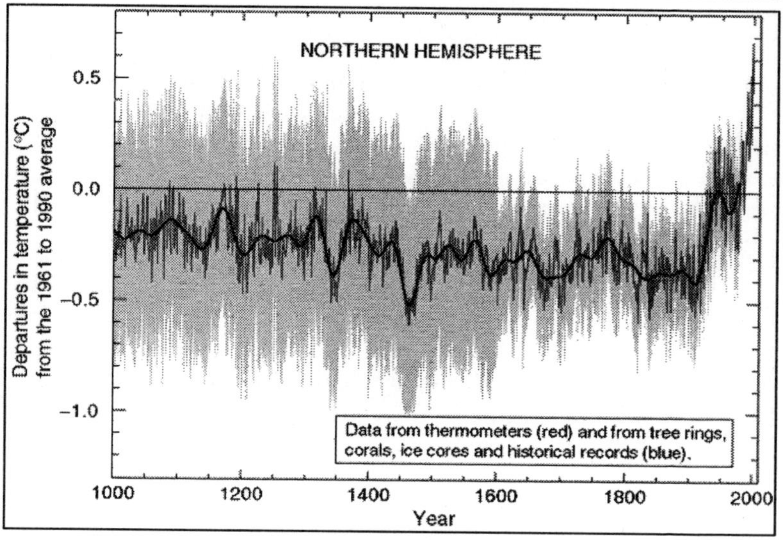

Figure 9.1: Hockey Stick Graph

Water

Rising global temperatures will lead to an intensification of the hydrological cycle, resulting in dryer dry seasons and wetter rainy seasons, and subsequently heightened risks of more extreme and frequent floods and drought. Changing climate will also have significant impacts on the availability of water, as well as the quality and quantity of water that is available and accessible. Melting glaciers will increase flood risk during the rainy season, and strongly reduce dry-season water supplies to one-sixth of the World's population.

Agriculture

Declining crop yields are likely to leave hundreds of millions without the ability to produce or purchase sufficient food supplies, especially in Africa. At mid to high latitudes, crop yields may increase for low levels of change in temperature, but will decline at higher levels of temperature change.

Ecosystems

Changing temperatures will cause ecosystems to shift – forests, land types and plant species will dieback in some areas as temperatures rise, but increase in other areas. However, in many cases, the pace of change in temperature may be too fast for ecosystems to adjust, resulting in the loss of forests and species.

Health

Higher temperatures expand the range of some dangerous vector-borne diseases, such as malaria, which already kills one million people annually, most of whom are children in the developing world. Further, heat waves associated with climate change, and increases in water borne diseases, will result in increased health problems.

Coastline

Melting ice and thermal expansion of oceans are the key factors driving sea level rise. In addition to exposing coastlines, where the majority of the human population lives, to greater erosion and flooding pressures, rising sea levels will also lead to salt water contamination of groundwater supplies, threatening the quality and quantity of fresh water access to large percentages of the population. For example, according to some estimates a 1 meter rise in sea level will displace 80 percent of the population of Guyana.

Climate Change and Hydrological Cycle

Introduction

Because much of the solar energy received by the Earth is used to drive the hydrological cycle, higher levels of solar energy trapped in the atmosphere will lead to an intensification of this cycle, resulting in changes in precipitation patterns. These changes will result in increased floods and drought, which will have significant impacts on the availability of freshwater. These impacts on freshwater will be further compounded by rising sea levels, and melting glaciers.

Floods and Drought

Warmer average global temperatures mean greater evaporation, with a warmer atmosphere able to hold more moisture aloft that can fall as precipitation, increasing the potential for flooding. There are a number of factors that contribute to flood risks implied by a warmer climate, including:

- ☆ More frequent wet spells in middle/high latitude winters;

- More intense mid-latitude storms;
- Increased frequency of extreme precipitation events;
- Increased magnitudes of precipitation events of high intensity; and
- Land use changes and surface degradation (for example, deforestation and urbanization).

The combination of more intense and frequent storms with land use changes is already proving to be deadly for the world's vulnerable populations. In 2004, a tropical storm that killed 18 people in the Dominican Republic resulted in the deaths of 3000 people across the border in Haiti; the difference has been attributed to deforestation, and the destruction of mangrove forests that are essential to coastal defense. Moreover, in South East Asia, areas with healthy mangrove and tree cover were significantly less likely to have experienced major damage as a result of the 2004 tsunami.

In drier regions, however, even a slight rise in temperatures will lead to greater loss of moisture, exacerbating drought and desertification. Drought will lead to decreased water availability and water quality for populations in many water-scarce regions, particularly in southern Africa, North Africa, Central America and central Asia. When less precipitation and higher temperatures occur simultaneously, the availability of water resources is reduced even further while evaporation is increased, leading to a vicious cycle.

In Sub-Saharan Africa, for example, long periods of drought are becoming more common and are predicted to become more widespread. While some farmers have been able to survive by selecting seed varieties based on changing conditions, poorer farmers have not been as able to adapt.

Rising Sea Level

Melting of ice and thermal expansion of oceans as a result of increased average temperatures will result in rising sea levels. Sea levels have already risen between 10-20 cm over pre-industrial averages and are expected to rise further by up to one meter during the 21 Century. Changes in sea level will increase coastal erosion, and flooding of wetlands and lowlands. It will also have a significant impact on the availability of freshwater:

- Groundwater resources in coastal plains are potentially vulnerable to salinization from rising sea levels, due to their low elevation and hydraulic gradient;
- Salt water intrusion in the lower reaches of the deltas will be exacerbated by predicted rises in sea levels; and
- Reduced protective capacity from extreme storms and floods, as higher sea levels provide a higher base for storm surges.

Melting Glaciers

Scientists are especially concerned about the Arctic, where the effects of climate change are being felt more quickly and severely than anywhere else on the planet. Arctic temperatures increased by about 5 degrees during the 20th Century – 10 times faster than the global average. Snow cover has declined by some 10 percent in the mid-high latitudes of the Northern Hemisphere since the late 1960s and, in the Russian Arctic, buildings are collapsing because of thawing permafrost under the foundations. The dramatic changes occurring in the Arctic are being felt around the world: melting glaciers contribute to rising sea levels, and also increased flash floods as river basins fill more quickly and with a greater volume of water, with the resulting impacts on freshwater availability. During dry seasons, water supplies are dramatically reduced due to reduced availability of glacial melt to fill streams and rivers. Furthermore, thawing permafrost in the Arctic is releasing methane, a gas which has a very high global warming impact. This will act as a further feedback mechanism, accelerating ongoing climate change.

Conclusion

Change in hydrological cycle, water quality and quantity implicitly are reflection of human behavior with environment. The wastes from industries are dissolved into the water resources available nearby without processing, thus polluting rivers and oceans. The recent oil spill in coast of Mumbai is perfect example of human error and deterioration of quality of water. Moreover, industrialization which is need of hour, is another important reason of change in hydrological cycle, because industrialization is affecting the environment, thus affecting climate and in all affecting the water.

We cannot have zero percent change neither in climate nor in hydrological cycle, because for development we need industrialization, *i.e.* a nation's development is directly proportional to industrialization. In a nutshell, hydrosphere is in peril and hydrological cycle is all set to be jeopardized.

Acknowledgement

It gives me immense pleasure to bring out this Chapter entitled" Impact of Climate Change on Water". I am very grateful to "WaterAid" (www.wateraid.org) whose Chapter on climate change and water resources gave me inspiration and comprehensive knowledge about these consequences. The author is grateful to WaterAid for providing informative support.

References

Climate change and water adaptation issues. Technical report No 2/2007. Cover Image. Published: 13 Feb 2007. www.eea.europa.eu/publications/technical_report_2007_2

Dr. Kathleen Miller's research on water and climate change is presented here for scientific or professional audiences.www.isse.ucar.edu/water_climate/impacts.html

IPCC Technical Paper Vi "Climate Change And Water".

IPCC. (2007) Climate change 2007: the physical science basis (summary for policy makers), IPCC.

Sea Level Change". University of Colorado at Boulder. http://sealevel.colorado.edu/documents.php. Retrieved 2009-07-21.

The United Nations Framework Convention on Climate Change". 21 March 1994. http://unfccc.int/essential_background/convention/background/items/1349.php.

Chapter 10

Global Climate Change and Salinity: Impact of Salinity on the Environment

Bhumija Kaphaliya and R.K. Srivastava

Availability of an adequate, safe water supply is critical to the health, economy, and environment of any country and its people. Current scientific research shows that climate change will have major effects on water cycle. Climate change is likely to increase the number of days of intense precipitation and the frequency of floods -ultimately to impart problems on water supply of the country. Thus climate-induced changes in the water cycle will affect the availability of water to meet growing demand. Climate change will also lead to the sea-level rise which will increase in the salinity of water in coastal areas. Sea-level rise over the last several decades has reportedly led to salinity intrusion and wetland loss in a number of coastal areas around the world. Wetlands play a key role in moderating climate change because their vegetation takes up carbon dioxide (CO_2) and acts as a large carbon sink. If wetlands are destroyed, the carbon (as CO_2) is no longer tied up and causes the CO_2 level in the atmosphere to rise. Salinity increase also causes an increase in soil salinity, especially when farmers irrigate their lands with saline water in arid and semi-arid areas because the available water in these areas is very saline. Thus we can say that the anticipated sea level rise due to climate change would produce salinity impacts in three fronts: surface water, groundwater and soil and vegetation. Increased soil salinity due to climate change would significantly reduce food grain production. Even at present, some parts of coastal lands are not being utilized for crop production, mostly due to soil salinity; and this situation would aggravate further under a climate change scenario. Rise in water temperature and salinity as a result of climate change will also alter fundamental ecological processes and the geographic distribution of aquatic species. By modeling exercise it was found that under the changed climate conditions, the index of aridity would increase in winter. Due to this, higher rates of capillary action from an increased rate of topsoil desiccation would accentuate the salinity problem. To minimize the problem of salinity due to climate change, salinity issue should be emphasized in integrated coastal zone management plans of the country. More study/ scientific research should be done for introducing salinity tolerant crop varieties. Drinking water crisis due to salinity should be incorporated in the National Water Policy. The global awareness should be raised to address vulnerability due to salinity as sea-level is rising.

Introduction

Earth is referred to as a blue planet because it appears blue when viewed from the space which is because 70 per cent of the earth surface is covered with water. However, 97 per cent of this water is saline, which cannot be used for drinking and irrigation purposes and only 3 per cent is the fresh water. The changing climatic condition due to anthropogenic activities is enhancing the salinity problem in water as well as in the soil day by day. Many arid and semi-arid parts of the world also face this salinity problem in the water as well as in the soil. The arid-semiarid regions of India, *i.e.* western state of Haryana, Punjab, Rajasthan and Gujarat have salinity problem. Salinity is the concentration of dissolved salts in water, usually expressed in "parts per thousand" (ppt). One potential impact of global warming and rise in sea level would be an increase in the salinity of estuaries, in a variety of ways. Sea-level rise will cause salinity to increase as seawater encroaches landward. Wind-induced changes in ocean currents may also lead to changes in the salinity distribution. Finally, changes in the climate of estuarine watersheds will change riverine freshwater input (Cayan and Peterson, 1993; Justic *et al.*, 1996) which might threaten drinking water and aquatic ecosystems. Sea-level rise over the last several decades has led to salinity intrusion and wetland loss in a number of coastal areas around the world. Wetlands play a key role in moderating climate change because their vegetation takes up carbon dioxide (CO_2) and acts as a large carbon sink. According to Banerjee and Gosh (1998), mangrove ecosystem of these wetlands commands the highest importance because of its biological productivity and specialized diversity. After coral reefs, the mangrove forest has the highest productivity among coastal wetlands (Sridhar and Curr, 2004).

Impact of Salinity due to Climate Change on Surface Water

Due to climate change and global warming salinity intrusion in the coastal rivers takes place. As sea level continues to raise, the associated effects of permanent inundation is likely to increase the salinity near coastal areas. According to a study done in Bangladesh, 5 ppt saline front will penetrate about 40 km inland for sea level rise of 88 cm which is going to affect the only fresh-water pocket of the Tetulia River in Meghna Estuary. A big chunk of the fresh- water zone that will be disappearing due to sea level rise near to the estuary will have a far reaching effect on the country's ecology and will extinct some of its endangered species (marked by IUCN) forever.

An Example of Salinity Intrusion

The clear evidence of increased saline intrusion in the coastal zones is in the coastal city of Khulna the main power station needs to collect fresh water to cools its boilers by sending a barge upstream to get freshwater. Over the last one decade the barge has to go further and further upstream to get suitably fresh water for the purpose. While there are other factors behind the salinity intrusion (such as the withdrawal of water at Farakka by India to divert water flows to Calcutta), the trend towards salinization in the coastal zone is very clear.

Impact of Salinity due to Climate Change on Wetland and Mangroves

India has a long coastal line of over 7500 km supporting vast habitats such as lagoons, backwaters, estuaries, coral reefs and mangrove swamps. Among them, the mangrove ecosystem has the highest importance because of its biological productivity and specialized diversity (Sandilyan *et al.*, 2010). Adeel and Pomeroy (2002) reported that mangrove ecosystem harbours unique and endemic species of plants and animals; moreover 90 per cent of all marine organisms spend some portion of their life cycle within the mangrove ecosystem. Due to increasing level of salinity such an important ecosystem is under severe pressure in recent years. The vital role of salinity in wetland habitats has been well established in several studies, especially in various saline habitats such as estuaries, swamps and lagoons (Nagarajan and Thiyagesan, 2006; Sumathi *et al.*, 2008; Sandilyan, 2009). Higher salinity reduces the biomass in *Bruguiera gymnorrhiza* (Naidoo, 1990). Koch and Snedaker (1997) documented the denaturing of terminal buds in *Rhizophora mangle* seedlings. Panda *et al.*, 2006, found that increased salinity reduces the growth of mangrove leaves. Munns and Termat (1986) reported that saline stress conditions due to higher salinity accelerate the leaf mortality rate and reduce the leaf primary production in mangrove plants. Lin and Sternberg (1993) found the negative effects on the photosynthesis mechanism in mangrove plants. Hoque *et al.* (2006) reported that a sudden reduction in mangrove forest cover affects the climax community seriously. They also concluded that an important species in the Sundarbans, 'sundari' (*Heritiera fomes*), faces a serious threat due to the salinity concentration prevalent in Bangladesh.

Effect of Salinity on Zooplanktons

Higher salinity would cause a profound impact on animals such as planktons, fungi, benthic forms, shrimps, crabs, fishes, water birds and other wild fauna living in and around the mangroves. According to Remane and Schlieper (1971), taxonomic richness in aquatic ecosystems tends to be low at salinities between 5 and 7psu. Schallenberg *et al.*, found that taxonomic richness declined dramatically with increasing salinity. However, the decline in actual taxonomic richness in the lake was less pronounced (though still significant) (Figure 10.1).

Water Salinity Removal

Very saline water is not recommended for the irrigation and drinking purposes so salinity removal from the water or management of saline water is very important to meet the growing demand of water. There are three basic categories of water purification

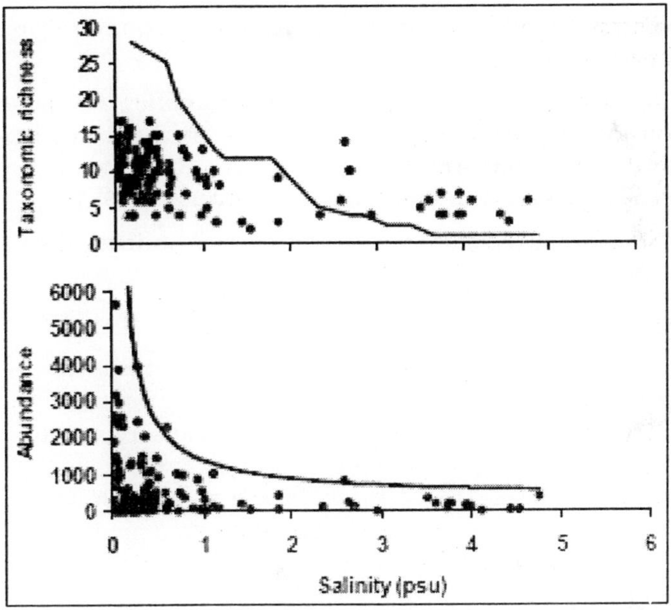

Figure 10.1: Decline in Zooplankton Taxonomic REichness and Abundance with Increasing Salinity. Line in the upper graph shows community structure based on abundance weighted averaging (*Source*: Schallenberg *et al.*)

technologies that are used for desalination: membrane technologies, distillation processes (thermal technologies), and chemical approaches. Some water purification plants use a combination of these technologies.

Membrane Technologies

According to American Water Works Association (1999), a membrane is a thin film of porous material that allows water molecules to pass through it, but simultaneously prevents the passage of larger and undesirable molecules such as viruses, bacteria, metals, and salts. These membranes can be made up of the polymeric materials that include cellulose, acetate, and nylon, and nonpolymeric materials such as ceramics, metals and composites. These membrane treatment processes are of two types. They can be either pressure-driven or electrical-driven technologies. Pressure-driven membrane technologies include reverse osmosis (RO), nanofiltration (NF), ultrafiltration, and microfiltration (Table 10.1) (Duranceau, 2001).

Reverse osmosis, and to some extent nanofiltration processes, are considered effective in salt removal. Electrical-driven membrane technologies that are effective with salt removal include electrodialysis (ED) and electrodialysis reversal (EDR).

Membrane technologies applicable to desalination are briefly described below.

Reverse Osmosis

Reverse Osmosis (RO) is a physical process that uses the osmosis phenomenon, *i.e.*, the osmotic pressure difference between the saltwater and the pure water to remove salts from water. In this process, a pressure greater than the osmotic pressure is applied on saltwater (feedwater) to reverse the flow, which results in pure water (freshwater) passing through the synthetic membrane pores separated from the salt. A concentrated salt solution is retained for disposal.

The RO process is effective for removing total dissolved solids (TDS) concentrations of up to 45,000 mg/L, which can be applied to desalinate both brackish water and seawater.

Nanofiltration

A nanofiltration (NF) membrane works similar to reverse osmosis except that with NF, less pressure is needed (70 and 140 psi) because of larger membrane pore size (0.05 m to 0.005 m). Nanofiltration can remove some total dissolved solids, but is often used to partially soften water and is successful at removing solids, as well as dissolved organic carbon. For low TDS brackish waters, NF may be used as a standalone treatment for removing salts.

Electrodialysis and Electrodialysis Reversal

Electrodialysis (ED) utilizes electromotive force applied to electrodes adjacent to both sides of a membrane to separate dissolved minerals in water. The separation of minerals occurs in individual membrane units called cell pairs. A cell pair consists of an anion transfer membrane, a cation transfer membrane, and two spacers. The complete assembly of cell pairs and electrodes is called the membrane stack (Figure 10.2). The number of cells within a stack

Table 10.1: Pressure-Driven Membrane Processes and their characteristics

Membrane Process	*Applied Pressure psi (kPa)*	*Minimum Particle Size Removed*	*Application (Type, average removal efficiency%)*
Microfiltration	4-70 (30-500)	0.1-3 µm	Particle/turbidity removal (>99%)
			Bacteria/protozoa removal (>99.99%)
Ultrafiltration	4-70 (30-500)	0.01-0.1 µm	Particle/turbidity removal (>99%)
			Bacteria/protozoa removal (>99.999%)
			TOC removal (<20%)
			Virus removal/(partial credit only)
Nanofiltration	70-140 (500-1000)	200-400 dalton	Turbidity removal (>99%)
			Color removal (>98%)
			TOC removal (DBP control) (>95%)
			Hardness removal (softening) (>90%)
			Synthetic organic contaminant (SOC) removal (500 daltons and up) (0-100%)
			Sulfate removal (>97%)
			Virus removal (>95%)
Hyperfiltration (Reverse osmosis)	140-700 (1000-5000)	50-200 dalton	Salinity removal (desalination) (>99%)
			Color and DOC removal (>97%)
			Radionuclide removal (not including radon) (>97%)
			Nitrate removal (85 to 95%)
			Pesticide/SOC removal (0 to 100%)
			Virus removal (>95%)
			As, CD, Cr, Pb, F removal (40 to >98%)

Source: Duranceau (2001)

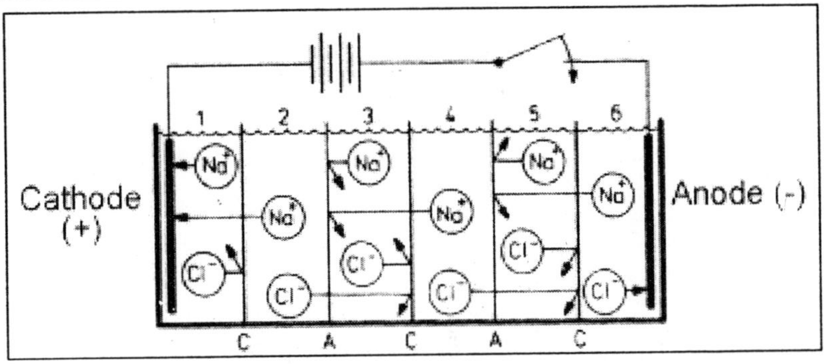

Figure 10.2: Electrodialysis Stack
(*Source*: Brunner, 1990)

varies depending on the system. The spacer material is important for distributing the water flow evenly across the membrane surface. The ED process is effective with salt removal from feedwater because the cathode attracts the sodium ions and the anode attracts the chloride ions. The required pressure is between 70 and 90 psi (Brunner, 1990). ED can remove 75 to 98 per cent of dissolve salids from feed water.

Thermal Technologies

Thermal technologies are the combination of evaporation and distillation processes. Modern thermal-based technologies are mostly developed as dual-purpose power and water desalination systems. These technologies are applied to desalination of seawater.

Chemical Process

Chemical approaches include processes such as ion exchange, which is considered impractical for treating waters with high levels of dissolved solids.

Ion Exchange Technologies

The ion-exchange system can best be described as the interchange of ions between a solid phase and a liquid phase surrounding the solid. Chemical resins (solid phase) are designed to exchange their ions with liquid phase (feedwater) ions, which purify the water. Resins can be made using naturally-occurring inorganic materials (such as zeolites) or synthetic materials. Ionexchange technologies applied to desalination are rather complex.

Table 10.2: Summary of Processes of Typical Thermal Technologies

Technology	*Brief Description: Advantages and Disadvantages*	*Feed Water*
Solar Distillation	A pond of saltwater with a clear lid takes advantage of solar heat. The saltwater evaporates and condenses on the lid. The brine stays in the pool lad condensation forms potable water	SW/BW
	Low energy costs, low material and equipment costs/requires large amount of land and direct sunlight, low productivities	
Multistage-Flash	Combination of many flashing stages. One flashing stage: Saltwater traveling through tubes is cooler than the vapor surrounding the tubes Heat exchange preheat the saltwater. The saltwater is emptied to the brine pool, where it evaporates and fills the vapor space that preheats the incoming saltwater. The vapor is condensed to form potable water, and the brine becomes the feed water for the next stage Proven reliable for years, can operate using waste thermal energy, can handle large capacities/requires highest amount of energy of all technologies	SW
Multiple Effect Evaporation (MEE)	Combination of many effects. One effect: Saltwater is sprayed overtop of hot lubes. It evaporates and the vapor is collected to run through the tubes in the next effect. As the cool saltwater is sprayed over the vapor filled tubes, the vapor condenses and is collected as potable water. The resulting brine collects in the bottom of each effect, and is either circulated to next effect or exited from system requires less energy than MSF, can operate using waste thermal energy can handle large capacities/high amounts of energy, scaling on tubing	
Thermal Vapor Compression (TVC)	Works as first effect of multiple effect evaporation. The steam jet ejector is used to compress the vapor for the tubes in the first effect. A condenser is responsible for condensing the vapor to the final product increases MEE performance ratio when combined with MEE	SW

Contd...

Table 10.2–*Contd...*

Technology	*Brief Description: Advantages and Disadvantages*	*Feed Water*
Mechanical Vapor Compression (MVC)	Works the same as thermal vapor compression except that mechanical compressors are used instead of steam jet ejectors meet needs in remote areas, transportable.	SW/BW
Adsorption Vapor Compression	Pressure differences occur between two tanks as a fluid mixture is transferred between them. This drives the heat exchange for evaporation and condensation of saltwater to form potable water	SW/BW
	Heat is released from an exothermic reaction between blending of feed water with a solution such as LiBr, which preheats the feed water that is sent to the evaporator	

Sources: El-Dessouky and Ettouney (2002), El-Dessouky *et al.* (2000)

Saltwater (feedwater) is passed over resin beads where salt ions from the saltwater are replaced for other ions. The process removes Na^+ and Cl^- ions from feedwater, thus producing potable water.

Conclusion

Thus we can say that the salinity removal from the water of the coastal regions as well as of arid and semiarid regions gets affected by the change in the global climate. By application of above mentioned salt removal technologies the people can get desalinated water for their daily requirements and for irrigation purposes. Further research is needed in this field to get cheap and cost effective salt removal technology, so that good quality water can be obtained at a lower price.

References

Adeel, Z. and Pomeroy, R., *Trees*, 2002, 16, 235–238.

American Water Works Association. 1999. *Manual of Water Supply Practices: Reverse Osmosis and Nanofiltration*. AWWA M46:173.

Banerjee, L. K. and Gosh, D., Species Diversity and Distribution of Mangroves in India. An Anthology of Indian Mangroves, ENVIS Publication, Annamalai University, 1998, pp. 20–24

Brunner, R. E. Electrodialysis. 1990. *Saline Water Processing*. Hans-Gunter Heitmann: VCH Verlagsgesellschaft, Federal Republic of Germany, 197-217.

Cayan, D. R., and Peterson, D. H. 1993. Spring climate and salinity in the San Francisco Bay estuary. Water Resour. Res. **29:** 293–303.

Duranceau, S. J. 2001. Reverse Osmosis and Nanofiltration Technology: Inorganic, Softening and Organic Control. Paper presented at the American Membrane Technology Association's Annual Symposium, Isle of Palms, S.C., August 5-8, 2001.

El-Dessouky, H. T., H. M. Ettouney, and F. Al-Juwayhel. 2000. Multiple Effect Evaporation-vapour Compression Desalination Processes. *Trans IchemE*, 78:662-676.

Hoque, M.A., Sarkar, M.S.K.A., Khan, S.A.K.U., Moral, M.A.H. and Khurram, A.K. M., *Res. J. Agricult. Biol. Sci.*, 2006, 2, 115–121.

Justic´, D., Rabalais, N. N. and Turner, R. E., 1996. Effects of climate change on hypoxia in coastal waters: A doubled CO2 scenario for the northern Gulf of Mexico. *Limnol. Oceanogr.* 41: 992–1003.

Koch, M.S. and Snedaker, S.C., *Aquat. Bot.*, 1997, 59, 87–98.

Lin, G. H. and Sternberg, L. D. S. L., *J. Exp. Bot.*, 1993, 44, 9–16.

Munns, R. and Termat, A., *Aust. J. Plant Physiol.*, 1986, 13, 143–160.

Nagarajan, R. and Thiyagesan, K., *Acta Zool. Sin.*, 2006, 52 (suppl.), 541–548.

Naidoo, G., *Aquat. Bot.*, 1990, 38, 209–219.

Panda, D., Dash, P. K., Dhal, N. K. and Rout, N. C., *J. Gen. Appl. Plant Physiol.*, 2006, 32, 175–180.

Remane, A, Schlieper, C. 1971. The biology of brackish waters. 2nd ed. Wiley Interscience.

Sandilyan, S., Habitat quality and waterbird utilization pattern of Pichavaram wetlands southern India. *Ph D thesis*, Bharathidasan University, Tiruchirapalli, India, 2009.

Sandilyan, S., Thiyagesan,K., Nagaraja, R. and Vencatesan, J. Salinity rise in Indian mangroves – a looming danger for coastal biodiversity. *Current Science*, vol. 98, no. 6, 25 March 2010.

Schallenberg, M., Hall, CJ, Burns, CW. Consequences of climate-induced salinity increases on zooplankton abundance and diversity in coastal lakes. *Mar. Ecol. Prog. Ser.* in press.

Sridhar, K.R., *Curr. Sci.*, 2004, **86**, 1586–1587.

Sumathi, T., Nagarajan, R. and Thiyagesan, K., *J. Sci. Trans. Environ. Technol.*, 2008, **2**, 9–17.

Chapter 11

Potential Impacts of Climate Change on Freshwater Resources: A Critical Review

Anil Kumar Shankhwar, Nanda Nautiyal, Bipin Kumar, Akanksha Rastogi, R.K. Srivastava and Vir Singh

Water is an essential, natural, inexorable, life-sustaining commodity for human beings. No one can imagine survival without it. However, day by day, water is going off from human access at an unprecedented rate. The root cause for dwindling potable water resources is over population, pollution and haphazard management of natural resources. Sustainable development could not be achieved until assurance of freshwater availability for future offspring. Climate change and global warming are also a driving factor to access freshwater in watercourse and human inhabitants. Accessibility of potable water for the increasing population is a growing challenge for scientific and government agencies in current era. The present manuscript attempts to determine the potential impacts of climate change on freshwater resources including responsible factors, assessment strategies, consequences of climate change and its management.

Introduction

The life on Earth is originated from sea and water plays pioneer role for evolutionary development of species, life forms and relatively complex molecules resulting in present status of biodiversity on Earth. Almost all physical, chemical and biological reactions require presence of water (Shankhwar, 2010). It originates, at least partly, from the Earth material itself (degassing of planetesimals during the early earth formation process) (Drake and Righter, 2002). Of all the planet's renewable resources, water has a unique place. It is essential for sustaining all forms of life, food production, economic development, and for general well being. It is impossible to substitute for most of its uses, difficult to de-pollute, expensive to transport, and it is truly a unique gift to mankind from nature.

Water is also one of the most manageable of the natural resources as it is capable of diversion, transport, storage, and recycling. All these properties impart to water its great utility for human beings (Kumar *et al.*, 2005). India receives annual precipitation of about 4000 km^3, including snowfall. Out of this, monsoon rainfall is of the order of 3000 km^3. Rainfall in India is dependent on the south-west and north-east monsoons, on shallow cyclonic depressions and disturbances and on local storms (Rao and Mamatha 2004).

Mall *et al.* (2006) evaluated the water resources of India which showed annual precipitation, available water resources and surface water etc. as in Table 11.1. It has recently explored the possible link between global warming and the intensification of the global water cycle by reviewing the current state of science regarding historical trends in hydrologic variables. The results of this analysis suggest that global warming may have indeed intensified the global water cycle over the course of the 20th century. The intensification of the global water cycle could affect the availability of water resources and may lead to an increase in the frequency and intensity of the tropic storms, floods, and droughts, and an amplification of the global warming through the water vapour feedback (Huntington, 2006).

Ugnar (1999) studied that the losses due to extreme events are increasing steeply specially in the last decade of the twentieth century. The stress on freshwater availability poses due to the consequences of water pollution. Among the various types of environmental pollution, water pollution is age-old problem, but it has gained an alarming dimension lately because of the problems of population increase, sewage disposal, industrial waste, radioactive waste, etc.

These factors have contributed so much to the pollution of water resources that nearly 70–80 per cent of rivers and streams all over the world carry polluted water (Karam and Nicell, 1997; Duran and Esposito, 2000). Figure 11.1 shows population increase and water availability (in cu. m./person/yr) decline in India (Mall *et al.* (2006).

Table 11.1: Water Resources of India

Annual precipitation	4000 b.cu.m.
Available water resources	1869
Utilizable	1122
Surface water (storage and diversion)	690
Groundwater (replenishable)	432
Present utilization (Surface water 63 per cent, groundwater 37 per cent)	605
Irrigation	501
Domestic	30
Industry, energy and other uses	74

Source: Mall *et al.*, 2006.

Continuation of this trend may have catastrophic impact on human health and environment. Therefore, effective means of solving this problem must be developed to preserve the quality of life for future generations. The management of freshwater is further complicated by the threat of climate change, which could potentially have a significant effect on the hydrologic cycle and hence on key parameters, such as rainfall and temperature. Global warming is generally expected to increase if current human emission of greenhouse gases continues unabated. Furthermore, in low-lying States, a rise in sea level may also cause flooding of low-lying areas, which may contaminate freshwater bodies, or increase salt water intrusion. Without adequate management, poor water quality or dwindling water supplies could set limits in the attainment of sustainable development. This review focuses on the impact of such climate change on global freshwater manifestation through different resources and it draws from the wide body of scientific evidence that has accumulated over the past few years.

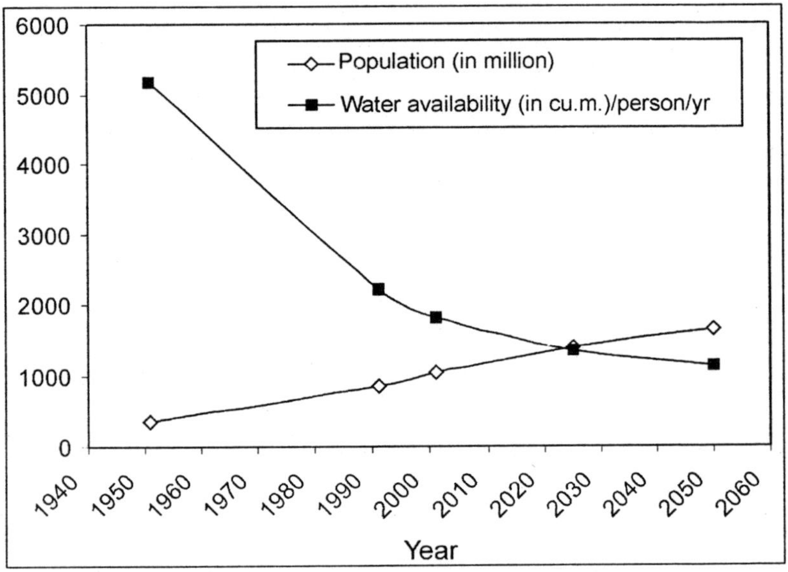

Figure 11.1: Observed and Projected Decline in per Capita Average Annual Freshwater Availability and Growth of Population from 1951 to 2050 (Mall *et al.*, 2006)

Key Factors Responsible for Global Climate Change

The climate of Earth is changing rapidly with human activities mostly responsible for this and many other global changes (Vitousek *et al.*, 1997; Sala *et al.*, 2000). Year to year deviations in the Indian weather and occurrence of climatic anomalies in respect of its four seasons are responsible due to (i) Cold wave, Fog, Snow storms and Avalanches; (ii) Hailstorm, Thunderstorm and Dust storms; (iii) Heat wave; (iv) Tropical cyclones and Tidal waves; (v) Floods, Heavy rain and Landslides, and (vi) Droughts (De *et. al.*, 2005).

India is a subtropical country with average annual rainfall of 117 cm, highly variable both in time and space. The heaviest rainfall of the order of 200–400 cm or even more occurs over northeast India and along the Western Ghats situated along the west coast of the peninsular India as shown in Figure 11.2 (Mall *et. al.*, 2006).

Weitzman (2007) reviewed stabilized greenhouse gas atmospheric concentrations at ≈k550 ppm of CO_2-equivalent (CO_2e). However, the current level is ≈430 ppm CO_2e, compared to ≈280

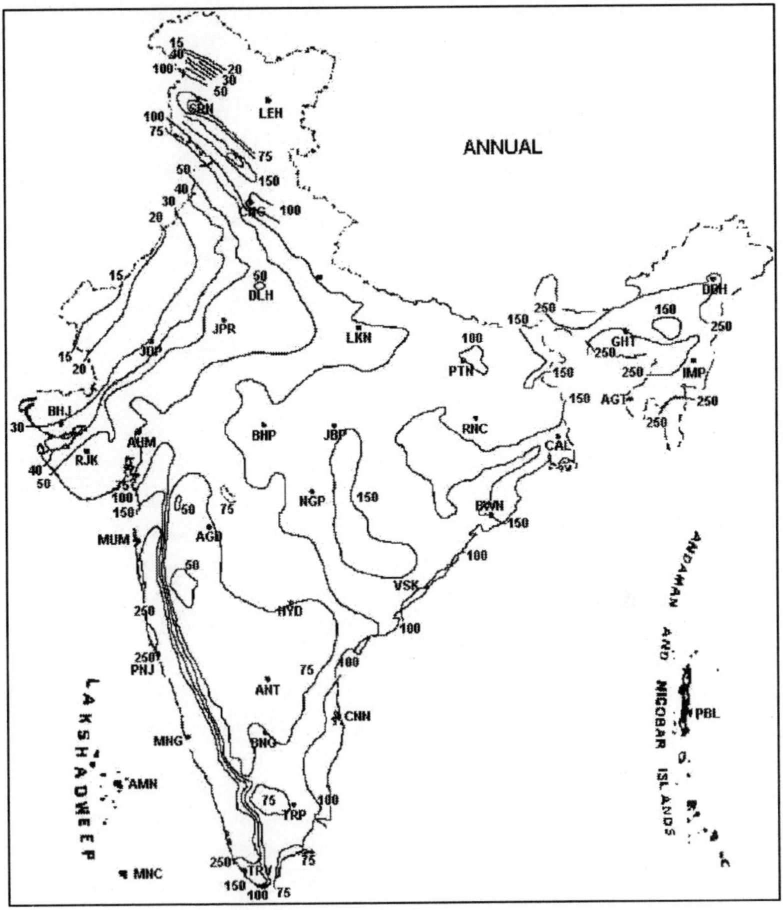

Figure 11.2: Normal Annual Rainfall (cm)
(*Source*: IMD)

ppm CO_2e before the industrial revolution. This would make temperatures a hundred years from now be at $E[\Delta T] \approx 2$°C and would (hopefully) stabilize future temperatures permanently thereafter at ΔTdeepak3°C. By contrast, along the more gradual majoritarian optimal trajectories CO_2e concentrations a century from now are >600 ppm and $E[\Delta T] \approx 2.5$°C with temperatures expected to continue rising to well above $E[\Delta T] \approx 3$°C after year 2105.

Beyond scarcity, water security is also about risk and vulnerability. One such risk is that of projected climate change impacts. The most detailed and sophisticated planning methods is use in the water sector treat weather as an uncertain but stationary process. In other words, climate is assumed to be fixed. However, increased climate variability is expected to alter the present hydrological resources and add pressure on the availability of future water resources in some parts of the world (Mukheibir 2010a).

Impact Identification Method

The analysis was done by Abler *et al.* (2002) for assessment of the potential impacts of climate change on water quality in the U.S. Chesapeake Bay region. The objectives were to answer the questions, *i.e.*, is climate change likely to improve or worsen agriculture's negative effects on water quality and at what extent; how sensitivity of its effects on water quality to economic responses by farmers, choice of baseline scenario, and to climate change impacts in other regions (other than Chesapeake Bay region U.S.) and countries? For answer a simulation model of maize production in twelve watersheds within the Chesapeake Bay Region was constructed with economic and watershed components linking climate to productivity. The results are split on the direction of change in water quality is an asymmetry in terms of the magnitude of change in nitrogen deliveries to the Chesapeake Bay. The largest fall in nitrogen deliveries were 33 per cent, while the largest rise was 57 per cent. However, a very limited picture of the overall consequences of climate-induced change appeared. Simulation model results indicate about the choice of baseline scenario that changes in nitrogen deliveries from maize production to the Chesapeake Bay differ significantly depending on weather. It also indicated a high degree of sensitivity that hinges on the ability of maize to productively use higher atmospheric levels of carbon dioxide.

Assessment of Potential Impacts on Freshwater Resources due to Climate Change

There is ample evidence that global climate change has already had a detectable effect on terrestrial to tropical marine environments (Walther *et al.*, 2002). However, impact assessment of climate change on freshwater resources is a crucial aspect to secure the future for potable water management perspective. The responses of both flora and fauna span an array of ecosystems and organizational hierarchies, from the species to the community levels. Climate change is a permanent change in the pattern and trend of weather over a long period of time expressed as a variation in the mean weather conditions, the likelihood of extreme conditions in a specific region or across the whole planet. Climate change can be attributed to either natural or anthropogenic drivers. Of most concern in the anthropogenic drivers is the increase in carbon dioxide (CO_2) levels due to emissions from fossil fuel combustion (brought about by the staggering global population growth and rising energy needs), followed by aerosols (particulate matter in the atmosphere) and land use; deforestation and agriculture (IPCC, 2007a).

Impact of Climate Change on Precipitation and Antarctic Region

In the context of climate change, Convey and Smith (2006) focused on three elements as fundamental to the biology of Antarctic terrestrial organisms, temperature, water and solar irradiance. Antarctic terrestrial biota is generally limited by the inexorably linked environmental factors of low summer temperature and lack of available water. However, in parts of the Antarctic, both these factors are changing rapidly on contemporary timescales. The availability of liquid water may be even more important than temperature to biological activity in Antarctic terrestrial habitats (Kennedy, 1993; Somme, 1995 and Block, 1996). Water availability is influenced directly by precipitation events, and also by release from seasonal snow cover and glaciers (*i.e.*, the timing can be separated from the original precipitation event). Increased precipitation is predicted in this region (Budd and Simmonds, 1991), with some documentary evidence published (Turner *et al.*, 1997). The decreasing trends in precipitation, more particularly in the sub-Antarctic (Frenot *et al.*, 1997; Bergstrom and Chown 1999), but also from the maritime Antarctic South Orkney is (Noon *et al.*, 2001), highlighting the importance of understanding trends at the appropriate (local) scale. Rapid decreases in glacial extent, and loss of smaller 'permanent' snow banks, may be both in the short-term release increased quantities of liquid water into terrestrial habitats. Such decreases are documented along the Antarctic Peninsula and Scotia Arc (Smith, 1990; Gordon and Timmis, 1992; Fowbert and Smith, 1994; Pugh and Davenport, 1997; Fox and Cooper, 1998; Vaughan *et al.*, 2001; Quayle *et al.*, 2003). In the longer term, loss or seasonal exhaustion of previously permanent sources of water supply may itself provide another limitation to terrestrial habitats (Convey *et al.*, 2003).

Qualitative and Quantitative Potential Impact of Climate Change on Fresh Water at City-Scale

A variety of potential impacts (and potential benefits) of climate change on cities had been identified and a number of recent reviews had described these, including the IPCC 3rd and 4th Assessment Reports (Scott *et al.*, 2001, and Wilbanks *et al.*, 2007). City-scale assessments are likely to be strengthened as a number of potentially significant climate change impacts are either unique to urban areas or exacerbated in urban areas (Lindley *et al.*, 2006). Hunt and Watkiss (2011) provided literature on assessment of the state of the art in the quantification and valuation of climate risks at the city-scale (Figure 11.3). For example, "surface sealing" that inhibits rainwater percolation leads to stress on urban drainage systems. The figure considers differences in methodological approaches, particularly relating to quantification and quality of city-scale climate change impacts and adaptation.

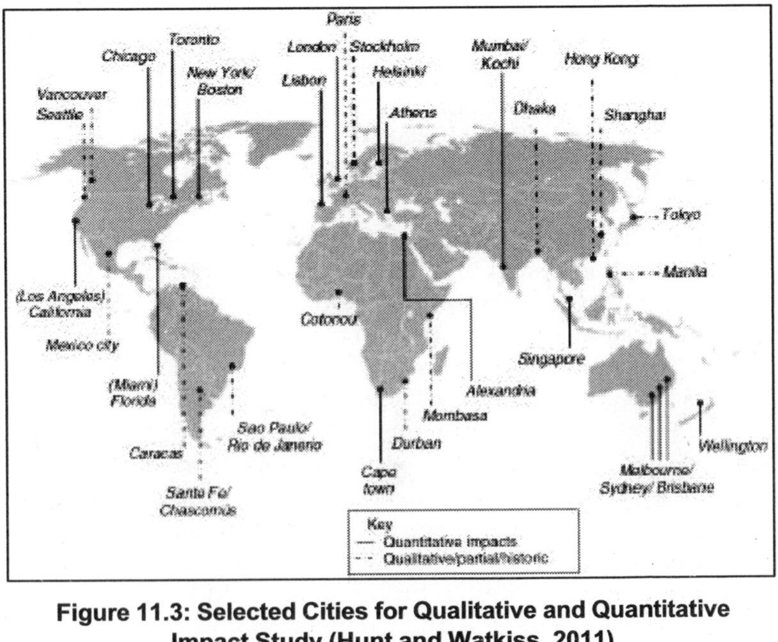

Figure 11.3: Selected Cities for Qualitative and Quantitative Impact Study (Hunt and Watkiss, 2011)

Effects of climate change on cities, which are directly related with water resources are likely to be:

- ☆ Effects of sea level rise on coastal cities (including the effects of storm surges);
- ☆ Effects of extreme events on built infrastructure (*e.g.*, from wind storms and storm surges, floods from heavy precipitation events, heat extremes and droughts);
- ☆ Effects on health (from heat and cold related mortality and morbidity, food and water borne diseases, vector borne diseases) arising from higher average temperatures and/ or extreme events;
- ☆ Effects on energy use (heating and cooling, energy for water); and
- ☆ Effects on water availability and resources.

Potential Impact on Agricultural Output

IPCC's Fourth Assessment Report (AR4) states that the benefits derived from climate change on fresh water systems, such as increased agricultural output and forest cover given that some areas will experience an increase in the annual runoff, are offset by the staggering negative impacts of climate change. These negative impacts include reduction in the value of the services provided by water resources especially where the runoff is projected to decline with climate change, increased precipitation variability such as intense flood events, flashier systems and seasonal runoff shifts on water supply and water quality (IPCC, 2007b). The Synthesis Report, the concluding document of the AR4 (page 49), stated that Climate change is expected to exacerbate current stresses on water resources from population growth and economic and land-use change, including urbanization. On a regional scale, mountain snow pack, glaciers and small ice caps play a crucial role in freshwater availability. Widespread mass losses from glaciers and reductions in snow cover over recent decades are projected to accelerate throughout the 21st century, reducing water availability, hydropower potential, and changing seasonality of flows in regions supplied by melt water from major mountain ranges (*e.g.*, Hindu-Kush, Himalayas, Andes), where more than one-sixth of the world population currently lives.

Impact on Hurricane

Although very recent studies have suggested a linkage between trends towards warming sea surface temperature and hurricane intensity in the Atlantic, it is also suggested that the empirical data to date do not consistently support an increase in the frequency or intensity of tropical storms and floods. However, there are substantial uncertainties regarding the trends in the hydrological and climatic variables due to the very variable quality of the data (both spatial and temporal), and to the regional variations and sometimes contradictory results. In spite of these uncertainties, the observed trend in most of the variables considered suggests that global warming may indeed have intensified the global hydrologic cycle over the course of the 20th century. Due to the potential effects of such a change on human well-being, it would be necessary to improve our capabilities to monitor and predict the consequences of changing the global water cycle. To that end, it would be necessary to improve our understanding of the complex feedbacks involving water and to manage longer periods of data collection at a larger scale Huntington, (2006).

Consecutive Climate Change Impacts on Different Freshwater Resources

Impacts on Glaciers

Glaciers are highly sensitive to changes in climate (Hoelzle *et al.*, 2003; Paul *et al.*, 2007; Raup *et al.*, 2007) highly susceptible to impact from even modest adjustments in local climatic conditions (Beniston, 2003). Mountain glaciers are often in close proximity to substantial human settlements, making changes to their size and appearance much more readily observable than changes to remote polar ice masses (Kaser, 2006). It is estimated that the Himalayan Mountains cover a surface area of permanent snow and ice in the region which is about 97,020 km^2 with 12,930 km^2 volume. In these mountains, 10 to 20 per cent of the total surface area is covered by glaciers, while an area ranging from 30 to 40 per cent has seasonal snow cover (Upadhyay, 1995; Bahadur, 1999). These glaciers provide snow and the glacial melt waters keep the Himalayan rivers perennial. Bahadur (1999) reported that a conservative estimate gives at least 500 km^3/yr as snow and ice melt water contributions to Himalayan streams, while Afford reports about 515 km^3/yr from the upper Himalayan mountains. The most useful facet of glacial run-off is the fact that glaciers release more water in a drought year and less water in a flood year and thus ensure water supply even during the lean years. The snow line and glacier boundaries are sensitive to changes in climatic conditions. Almost 67 per cent of the glaciers in the Himalayan mountain ranges have retreated in the past decade (Ageta and Kadota, 1992). According to IPCC (1998), available records suggest that the Gangotri glacier is retreating about 28 m per year. A warming is likely to increase the melting more rapidly than the accumulation. Glacial melt is expected to increase under changed climate conditions, which would lead to increased summer flows in some river systems for a few decades, followed by a reduction in flow as the glaciers disappear.

The additional geomorphological hazards, water resources vulnerabilities, and risk from sea-level rise that are a result of near universal glacial melting has global social, economic and geopolitical consequences. Kling *et al.* (2005) report that winters are shorter, average annual temperatures are rising, and the duration of lake ice cover is decreasing as air and water temperatures rise. An assessment report states that over the last 30 years the maximum amount of ice forming on each of the five lakes each year is decreasing.

Impacts on Wetland and Mangrove

Wetlands contain about 20 percent of all terrestrial carbon stocks so they are an important carbon sink. Covering just six percent of Earth's land surface, wetlands – including marshes, peat bogs, swamps, river deltas, mangroves, tundra, lagoons and river floodplains – contain an estimated 771 billion tons of greenhouse gases, both CO_2 and more potent methane, an amount in CO_2 equivalent comparable to the carbon content of today's entire atmosphere (Leahy, 2010). It is estimated that climate change may result in a loss of up to 85 percent of wetlands in the future. That loss would also release enough carbon and methane to almost certainly tip the climate into an era of extreme and rapid change as experts believe. Under natural conditions, coastal wetlands adjust to rising seas and changes in local storm patterns, but climate changes and human activities that alter natural conditions disrupt wetland hydrology, biogeochemical cycling, and other processes that sustain wetlands (Morris *et al.*, 2002).

Evaporation and Evapo-transpiration Impacts on Great Lake

Millerd (2011) examines the potential impacts of the lower water levels for the Great Lakes that may result from global climate change on Great Lakes international commercial navigation. Climate change will likely be manifested in the Great Lakes area by higher temperatures, possibly causing increased evaporation and evapotranspiration, lower runoff into rivers and lakes, higher lake temperatures; and reduced ice formation with shorter periods of ice cover. Rainstorms may be more intense and, due to higher temperatures, more precipitation may fall as rain rather than snow. A number of authors report that overland evapotranspiration may increase and total runoff to the lakes may be lower due to the higher temperatures. Also, runoff may peak earlier due to a lower snowpack, changing the seasonal distribution of water levels. Overall, the average steady-state supply of water to the lakes may decrease, resulting in lower lake levels and lower connecting channel flows. Chao (1999), Croley (1990), Easterling and Karl (2001), and Mortsch and Quinn (1996) concluded that any increases in precipitation are not expected to be sufficient to overcome the increased evaporation and evapotranspiration. Millerd (2011) examines the implications of lower water levels in the Great Lakes, but some predict that Great Lakes water levels may not fall or may even rise as a result of climate change. Manabe *et al.* (2004) and Milly *et al.* (2005) after modeling future world-wide changes in water availability due to climate change predict that run-off will increase in high latitude northern areas, resulting in higher levels for the Great Lakes. Manabe *et al.* (2004) simulations show an increase in the flow of the St. Lawrence River. Kutzbach *et al.* (2005) report on simulations from eight climate change models using two greenhouse gas emission scenarios. Averaging the model results shows that the Great Lakes will experience increased moisture but the actual change in the annual water balance was in doubt due to variations between models in precipitation and evaporation. There is evidence that the climate of the Great Lakes is already changing. Kling *et al.* (2005) report that winters are shorter, average annual temperatures are rising, and the duration of lake ice cover is decreasing as air and water temperatures rise.

Millerd (2011) concluded that the expected higher temperatures of climate change are predicted to increase evaporation, lower runoff, reduce ice formation, and raise surface water temperatures in the Great Lakes, possibly resulting in a fall in lake levels. Increases in precipitation will not be sufficient to completely offset the reduction in lake levels. As well as Climate change is also expected to increase the threat of ecological and economic damage from aquatic invasive species.

Impacts on Biodiversity of Freshwater Course

The uncertainties about representation of the complete impact of climate on water body's organisms are the greatest challenge in evaluating the future response of freshwater biodiversity to climate change (Allan *et al.*, 2005). For example, declines in levels of one metre or more were projected for the Great Lakes of Canada and the USA in the mid-nineties (Mortsch and Quinn, 1996; Mortsch *et al.*, 2000). But subsequent models have predicted smaller decreases or even increases in water levels (Lofgren *et al.*, 2002). Many species are declining (Pimm *et al.*, 1995; Brown *et al.*, 2007; Durance and Ormerod, 2007) whilst exotic species are appearing where they did not occur before (Lodge, 1993; Dukes and Mooney, 1999) and relationships between them are changing, with consequences throughout the ecosystem (Winder and Schindler, 2004).

Impacts of Climate Change on Indian Lake

The regular and periodic changes in the climate synchronized with season are ultimately reflected in the environmental parameters also, which in turn have a direct or indirect influence over the planktonic population. The seasonal distribution, abiotic and biotic processes affect the nutrient cycle of different coastal environments (Choudhury and Panigrahy, 1991).

Assessment of water resource quality of any region is an important aspect of developmental activities of the region, because rivers, lakes and man-made reservoirs are used for water supply to domestic, industrial, agricultural and fish culture (Jakher and Rawat, 2003). Global warming and climatic change are responsible for significant deleterious impact on lake water quality and which can be determined on the basis of physico-chemical and biological quality status of that lake ecosystem. Nainital Lake which is extended on Himalayan Mountain foothill also influenced at significant scale as per the trend of deteriorated water qualities with the increasing water temperature (Shankhwar, 2010). Temperature can have distinctive impact on several parameters in the water. Thermal stratification which occurs in the reservoirs as a result of water's temperature dependent density have a direct impact on levels of dissolved oxygen in the water column. Dissolved oxygen can provide information about the overall functioning and productivity of the reservoir. Oxygen is required for respiration by most aquatic organisms (Gastarova *et al.*, 2008). Thermal stratification begins in early spring when rising air temperatures warm up surface water. As air temperatures increase in spring and summer, reservoir water also warms up. Surface water warms more rapidly than deeper water because the warmer surface water is lighter than the colder, deeper water. These lakes separate into three distinct thermal layers. The top, warm layer is normally called the epilimnion, and the lower, cooler layer is the hypolimnion. Mediterranean deep lakes will suffer lower water levels, raised salinities (Green *et al.*, 1996) and elimination of much of their littoral zone with a major reduction in biodiversity. They have responded to warmer winter and early spring temperatures by shallower mixing during winter and earlier spring thermal stratification, and increased density and biomass of plankton. Zooplankton growth, however, also occurs earlier and there may be earlier suppression of phytoplankton by zooplankton grazing (Straile, 2000; Anneville *et al.*, 2005) Earlier warming and higher average water temperature often lead to deeper and longer-lasting stratification. This results in a longer-lasting dark hypolimnion, (Manca *et al.*, 2007). Warming may enhance the symptoms of eutrophication by increasing the rate of release of phosphorus from sediments (McKee *et al.*, 2003), increasing phytoplankton growth rates (Van Donk and Kilham, 1990; Reynolds, 1997; Howard and Easthope, 2002) and encouraging the growth of exotic plant species that displace native species (McKee *et al.*, 2002), not least floating species (Feuchtmayr *et al.*, 2009), some of them such as *Salvinia* and *Pistia* spp. causing major problems. Small ponds may suffer more frequent drying out in warmer summers and in the Mediterranean zone many will disappear (Blondel and Aronson, 1999). Lakes have the problem (and sometimes advantage) of greater isolation than river systems and although flying insects will readily colonize a new lake, migration of species from the south to replace those that do not survive warming will be impeded by isolation (Reist *et al.*, 2006).

Table 11.2: Some Current Studies Related with Climate Change Impact on Water Resources

Location	*Study Nature*	*Type of Study*
Helsinki (Europe)	Climate change in urban planning (flooding) VTT (2008)	Quantitative
Boston (North America)	Climate's long-term impacts on Metro Boston. Transport, Energy, Health (all quantitative) and Water (valuation) (Kirshen *et al.*, 2006).	Quantitative/ valuation
California (Los Angeles) (North America)	Heat mortality (quantitative), water availability and ecosystems under future climate (Hayhoe *et al.*, 2004) Cayan *et al.* (2006). Electricity. Miller *et al.* (2007)	Quantitative
Florida (Miami) (North America)	Sea level rise, hurricane damage, energy for cooling and tourism (Stanton and Ackerman, 2007)	Quantitative/ valuation
Dhaka (Bangladesh)	Flood Management and Vulnerability of Dhaka City (Huq and Alam, 2003). Alam and Rabbani (2007). Climate change induced flooding and air quality impacts (Alam and Rabbani, 2007). Historic impacts and qualitative future impacts.	Qualitative
London and New York	Assessment of climate risks and adaptation (Hunt and Watkiss, 2011)	Quantitative
United States or Canada	Potential impacts of the lower water levels for the Great Lakes including international commercial navigation (Millerd, 2011).	Quantitative

Other Consequences of climate change Related with Freshwater Ecosystem

It is clear that global warming has started negatively affecting a wide variety of organisms world-wide. Extinctions have started, and many organisms are being pushed closer to extinction or local extermination as a direct or indirect result of climate change (Kannan, 2009).

With the melting of large areas of the Antarctic ice shelf, more liquid and warmer water is available for colonization, and therefore, several Antarctic plants and invertebrates have shown striking distributional changes (Smith, 2001).

One of the most destructive effects in association with global warming is evident in the coral reefs of the world's oceans (Glynn 1991; Hoegh-Guldberg, 1999; Mumby *et al.*, 2001).

Global warming could directly affect human fishery industry. Variations in atmospheric air circulation over the Bering Sea impede the movement of juveniles away from adults of a commercially and ecologically vital species of fish, the Walleyed Pollock (*Theragra chalcogramma*) (Cushing, 1995).

Costs and Benefits of Climate Change in order to determine the impact of climate change on the water sector, projections have to be made about water use under a scenario of no climate change. This scenario was determined in this study, and it was assumed that the mean monthly temperature and rainfall of the baseline period existed for all years from 2007 to 2099.When the relative changes by sector for each country are applied to actual water demand in 2006, the total change in water demand. Overall, relative to 2006, the total demand for water in the Caribbean is expected to fall by 2030 by 11.3 per cent to approximately 12,967 million cubic meters. This is due to the expected fall in agricultural water demand by approximately 36 per cent in that period. However, by 2050, total water demand for the Caribbean will again exceed the 2006 level by approximately 4 per cent to 14,896.33 106 m^3. By 2100, water demand will increase almost fivefold to approximately 69,233.69 106 m^3. Climate change may also have other impacts on commercial navigation. Extreme weather events may become more intense and frequent, forcing vessels to delay voyages and vessels may be forced to control engine exhaust emissions or buy emission credits under a cap and trade system, an additional expense.

Sustainable Water Management in Urban Areas:

- ☆ Return of water after advanced purification to the reservoir or aquifer feeding the central water works (concept planned for Singapore) (Wah , 2007).
- ☆ Decentralized infiltration of roof and road runoff into the aquifer after proper purification; Athanasiadis *et.al.* (2007).
- ☆ Direct reuse of water, nutrients, and bio-solids in decentralized units after source separation (*e.g.*, urine, feces, and grey water) (Wilderer 2004; Cornel 2007).

Gorshkov *et al.* (2000) led to the conclusion that the continuing presence of liquid water on Earth is the result of a very complex interrelationship between biological and physico-chemical processes. Moreover, it appears that in the millennia following the appearance of water on Earth evolution, proliferation, and diversification of phototrophic and heterotrophic bacteria as well as primitive forms of higher biota made it possible that the original gas composition of the atmosphere changed, and eventually reached the state as we know it today. Sometime minute amount of contaminant is enough to change or deteriorate the water quality drastically (European Union determined permissible limit for fat in eggs as 3 ppb of dioxins), for unsafe drinking water (*e.g.* Dioxin deteriorates plastic bottle water quality) and yet it is avoidable also Shankhwar and Srivastava (2011).

The debate about measures to be taken in order to combat global warming and climate change is currently almost exclusively focused on the reduction of CO_2 emissions. It is advisable to add a new dimension to this debate, *i.e.*, strengthening the self-regulatory capacity of the Earth system. Major efforts need to be made to preserve this capacity, particularly with the aim of avoiding further increases in the mass of water vapour in the atmosphere. Global warming is not only a CO_2 problem; it is also a water problem. On a local scale, development of innovative concepts and methods to manage urban areas is required. Here we are facing rapid growth in the demand for water and food, and rapid deterioration of water resources, both quantitatively and qualitatively (Wilderer, 2009).

References

Abler, D.; Shortle, J.; Carmichael, J. and Horan, R. 2002. Climate Change, Agriculture, and water quality In the Chesapeake Bay Region, *Climatic Change.* 55: 339–359.

Afford, D. 1992. ICIMOD Occasional Paper No. 18, Kathmandu, p. 68.

Ageta, Y. and Kadota, T. 1992. *Ann. Glaciol.* 16: 89–94.

Allan, J.D.; Palmer, M. and Poff, N.L. 2005. Climate change and freshwater ecosystems. In: *Climate Change and Biodiversity.* Lovejoy, T.E. and Hannah L. (eds). New Haven Connecticut, Yale University. Press, pp. 272–90.

Anneville, O.; Gammeter, S. and Straile, D. 2005. Phosphorus decrease and climate variability: mediators of synchrony in phytoplankton changes among European peri-alpine lakes. *Freshwater Biology.* 50: 1731-1746.

Athanasiadis, K.; Helmreich, B. and Horn, H. 2007. On-site infiltration of a copper roof runoff: The role of clinoptilolite as an artificial barrier material. *Water Research,* 41(15): 3251–3258.

Bahadur, J. 1999. *The Himalayan Environment* (eds Dash, S. K. and Bahadur, J.), New Age International (P) Ltd, New Delhi, 258–267.

Beniston, M. 2003. Climatic Change in Mountain Regions: A Review of Possible Impacts. *Climatic Change.* 59 (1-2): 5-31.

Bergstrom D.M. and Chown S.L. 1999. Life at the front: history, ecology and change on southern ocean islands. *Trends Ecol. Evol.* 14: 472 –476.

Block W. 1996. Cold or drought – the lesser of two evils for terrestrial arthropods? *Eur. J. Entomol.* 93: 325–339.

Blondel, J. and Aronson, J. 1999. *Biology and Wildlife of the Mediterranean Region.* Oxford University Press, Oxford.

Brown, L.E., Hannah, D.M. and Milner, A.M. 2007. Vulnerability of alpine stream biodiversity to shrinking glaciers and snowpacks. *Global Change Biology* 13: 958-966.

Budd, W.F. and Simmonds, I. 1991. The impact of global warming on the Antarctic mass balance and global sea level. In: Weller

G., Wilson C.L. and Severin B.A.B. (eds), *Proceedings of the International Conference on the Role of Polar regions in Global Change.*

Chao, P. 1999. Great lakes water resources: climate change impact analysis with transient GCM scenarios. *J Am Water Resour Assoc.* 35(6):1499–1507.

Choudhury, S.B. and Panigrahy, R.C. 1991. Seasonal distribution and behaviour of nutrients in the creek and coastal waters of Gopalpur, east coast of India. *Mahasagar-Bull. Nat. Inst. Oceanogr.* 24:81-88.

Convey P., Block W. and Peat H.J. 2003. Soil arthropods as indicators of water stress in Antarctic terrestrial habitats?. *Global Change Biol.* 9:1718–1730.

Convey, P. and Smith, R.I.L. 2006. Responses of terrestrial Antarctic ecosystems to climate change. *Plant Ecology* 182:1–10.

Cornel, P. 2007. Semi-centralized supply and treatment systems for fast growing areas. In: Huber H H, Wilderer P A, Paris S, (eds.) *Water Supply and Sanitation for All.* Hans Huber AG Berching, pp. 211–222.

Croley, T.E. 1990. Laurentian Great Lakes double-CO_2 climate change hydrological impacts. *Climate Change.* 17(1): 27–47.

Cushing, D.H. 1995. *Population Production and Regulation in the Sea: A Fisheries Perspective.* Cambridge University Press, U.K.

De, U.S.; Dube, R.K. and Rao, G.S.P. 2005. Extreme Weather Events over India in the last 100 years. *J. Ind. Geophys. Union* 9(3): 173-187.

DENR. 1999. The Philippines' Initial National Communication on Climate Change to the UNFCCC.

Drake M J and Righter, K. 2002. Determining the composition of the Earth. *Nature* 416: 39–44.

Dukes, J.S. and Mooney, H.A. 1999. Does global change increase the success of biological invaders? *Trends in Ecology and Evolution* 14: 135-139.

Duran, N., and Esposito, E. 2000. Potential applications of oxidative enzymes and phenoloxidase-like compounds in wastewater and soil treatment: A review.

Durance, I. and Ormerod, S.J. 2007. Climate change effects on upland stream macroinvertebrates over a 25-year period. *Global Change Biology* 13: 942-957.

Easterling, D.R. and Karl, T.R. 2001. Potential consequences of climate change and variability for the Midwestern United States In National Assessment Synthesis Team (eds) *Climate Change Impacts on the United States–Foundation Report*. Cambridge University Press, Cambridge UK

Feuchtmayr, H., Moran, R., Hatton, K., Connor, L., Heyes, T., Moss, B., Harvey, I. and Atkinson, D. 2009. Global warming and eutrophication: effects on water chemistry and autotrophic communities in experimental, hypertrophic, shallow Lake Mesocosms. *Journal of Applied Ecology* 46, 713-723.

Fowbert J.A. and Smith R.I.L. 1994. Rapid population increase in native vascular plants in the Argentine Islands, Antarctic Peninsula. *Arct. Alpine Res.* 26: 290–296.

Fox, A.J. and Cooper, A.P.R. 1998. Climate-change indicators from archival aerial photography of the Antarctic Peninsula. *Ann. Glaciol.* 27: 636–642.

Frenot, Y., Gloaguen, J.C. and Trehen P. 1997. Climate change in Kerguelen Islands and colonization of recently deglaciated areas by *Poa kerguelensis* and *P. annua*. In: Battaglia B., Valencia J. and Walton D.W.H. (eds), *Antarctic Communities: Species, Structure and Survival*. Cambridge University Press, Cambridge.

Glynn, P.W. 1991. Coral reef bleaching in the 1980s and possible connections with global warming. *Trends in Ecology and Evolution* 6: 175-179.

Gordon J.E. and Timmis R.J. 1992. Glacier fluctuations on South Georgia during the 1970s and early 1980s. *Antarct. Sci.* 4: 215–226.

Gorshkov, V.G., Gorshkov, V.V. and Makarieva, A.M. 2000. Biotic Regulation of the Environment: Key Issue of Global Change. Berlin: Springer.

Green, A.J., Fox, A.D., Hilton, G., Hughes, B., Yarar, M. and Salathé, T. 1996. Threats to Burdur Lake ecosystem, Turkey and its waterbirds, particularly the white-headed duck *Oxyura leucocephala*. *Biological Conservation* 76: 241-252.

Hoegh-Guldberg, O. 1999. Climate change, coral bleaching and the future of the world's coral reefs. *Marine and Freshwater Research* 50: 839-866.

Hoelzle, M., Haeberli, W., Dischl, M. and Peschke, W. 2003. Secular glacier mass balances derived from cumulative glacier length changes. *Global and Planetary Change* 36: 295-306.

Howard, A. and Easthope, M.P. 2002. Application of a model to predict cyanobacterial growth patterns in response to climatic change at Farmoor Reservoir, Oxfordshire, UK. *Science of the Total Environment* 282, 459-469.

Hunt, A. and Watkiss, P. 2011. Climate change impacts and adaptation in cities: a review of the literature, *Climatic Change* 104:13–49.

Huntington, T.G. 2006. Evidence for intensification of the global water cycle: Review and synthesis, *Journal of Hydrology* 319: 83–95.

IPCC. 1998. *Special Report of IPCC Working Group II* (eds Watson, R. T. *et al.*), Intergovernmental Panel on Climate Change, Cambridge University Press, Cambridge, UK, 517.

IPCC. 2007a. *Summary for Policymakers in Climate Change 2007: The Physical Science Basis. Contribution of Working Group I to the Fourth Assessment Report of the Intergovernmental Panel on Climate Change.* Cambridge University Press, United Kingdom and New York, USA.

IPCC. 2007b. *Climate Change 2007: Synthesis Report.* Formally agreed statement of the IPCC concerning key findings and uncertainties contained in the Working Groups contributions to the Fourth Assessment Report. Pachauri, R.K. and Reisinger, A. (Eds.) IPCC, Geneva, Switzerland 104.

Kannan, R. and James, D.A. 2009. Effects of climate change on global biodiversity: a review of key literature *Tropical Ecology* 50(1): 31-39.

Karam, J., and Nicell, J.A. 1997. Potential applications of enzymes in waste treatment. *J. Chem. Technol. Biotechnol.* 69: 141–147.

Kaser, G. 2006. Mountain glaciers. In *Glacier Science and Environmental Change,* ed. P.G. Knight, Malden, MA: Blackwell Publishing. 268-271.

Kennedy, A.D. 1993. Water as a limiting factor in the Antarctic terrestrial environment: a biogeographical synthesis. *Arct. Alpine Res.* 25: 308–315.

Kumar, R., Singh, R.D. and Sharma K.D. 2005. Water Resources of India, *Current Science*, 89: 5.

Leahy, S. 2010. Wetlands loss fuelling CO_2 feedback loop. Retrieved July 5, 2010 from IPS Inter Press Service News Agency (IPS) Website at http://ipsnews.net/news.asp?idnews=43254

Lindley, S.J., Handley, J.F., Theuray, N., Peet, E. and McEvoy, D. 2006. Adaptation strategies for climate change in the urban environment: assessing climate change risks in UK urban areas. *J Risk Res* 9(5):543–568.

Lodge, D.M. 1993. Biological invasions: lessons for ecology. *Trends in Ecology and Evolution* 8: 133-137.

Lofgren, B. M., Quinn, F. H., Clites, A. H., Assel, R. A., Eberhardt, A. J., and Luukkonen, C. L. 2002. Evaluation of potential impacts on Great Lakes water resources based on climate scenarios of two GCMs. *Journal of Great Lakes Research*, 28: 537–54.

Mall, R.K., Gupta, A., Singh, R., Singh, R.S. and Rathore L.S. 2006. Water Resources and Climate Change: An Indian perspective, *Current Science*, 90: 12.

Manca, M.M., Portogallo, M. and Brown, M.E. 2007. Shifts in phenology of *Bythrotrephes longimanus* and its modern success in Lake Maggiore as a result of changes in climate and trophy. *Journal of Plankton Research* 29: 515-525.

McKee, D., Atkinson, D., Collings, S.E., Eaton, J.W., Gill, A.B., Harvey, I., Hatton, K., Heyes, T., Wilson, D. and Moss, B. 2003. Response of freshwater microcosm communities to nutrients, fish, and elevated temperature during winter and summer. *Limnology and Oceanography* 48, 707-722.

McKee, D., Hatton, K., Eaton, J.W., Atkinson, D., Atherton, A., Harvey, I., and Moss, B. 2002. Effects of simulated climate warming on macrophytes in freshwater microcosm communities. *Aquatic Botany* 74, 71-83.

Millerd, F. 2011. The potential impact of climate change on Great Lakes international shipping, *Climatic Change* 104: 629–652.

Morris, J.T., Sundareshwar, P.V., Nietch, C.T., Kjerfve, B. and Cahoon D.R. 2002. Response of coastal wetlands to rising sea level. *Ecology* 83: 2869-2877.

Mortsch, L. and Quinn F. 1996. Climate change scenarios for Great Lakes Basin ecosystem studies. *Limnology and Oceanography,* 41(5): 903–11.

Mortsch, L., Hengeveld, H., Lister, M., Lofgren, B., Quinn, F., Slivitzky, M., and Wenger, L. 2000. Climate change impacts on the hydrology of the Great Lakes-St. Lawrence System. *Canadian Water Resources Journal,* 25(2): 153–79.

Mukheibir, P. 2010b. The potential economic impact of climate change on equitable water access in small towns: a South African case study. *International Journal of Water,* 5(3): 223–245.

Mukheibir, P. 2010a. Water Access, Water Scarcity, and Climate Change, *Environmental Management,* 45: 1027–1039.

Mumby, P.J., Chrisolm, J.R.M. Edwards, A.J. Clark, S.D., Roark, E.B., Anrefouet, E. and Jaubert, J. 2001. Unprecedented bleaching-induced mortality in Porites species at Rangiroa Atoll, French Polynesia. *Marine Biology* 139: 183-189.

Noon P.E., Birks, H.J.B., Jones, V.J. and Ellis-Evans, J.C. 2001. Quantitative models for reconstructing catchment ice-extent using physical-chemical characteristics of lake sediments. *J.Paleolimnol.* 25: 375–392.

Paul, F., Kääb, A. and Haeberli. W. 2007. Recent glacier changes in the Alps observed by satellite: Consequences for future monitoring strategies. *Global and Planetary Change* 56: 111-122.

Pimm, S.L., Russell, G.J., Gittleman, J.L. and Brooks, T.M. 1995. The future of biodiversity. *Science* 269: 347-350.

Pugh P.J.A. and Davenport J. 1997. Colonisation vs. disturbance: the effects of sustained ice-scouring on intertidal communities *J. Exp. Mar. Biol. Ecol.* 210: 1 –21.

Quayle, W.C, Convey, P., Peck, L.S., Ellis-Evans, J.C., Butler, H.G. and Peat, H.J. 2003. Ecological responses of maritime Antarctic lakes to regional climate change. In: Domack E., Burnett A., Leventer A., Convey P., Kirby M. and Bindschadler R. (eds), *Antarctic Peninsula Climate Variability: Historical and*

Palaeoenvironmental Perspectives., Antarctic Research Series. American Geophysical Union, Washington D.C, 79: 159–170.

Rao, S. M. and Mamatha, P. 2004. Water quality in sustainable water management. *Curr. Sci.* 87: 942-946.

Raup, B., Racoviteanu, A. Khalsa, S.J.S., Helm, C., Armstrong, R. and Arnaud, Y. 2007. The GLIMS geospatial glacier database: A new tool for studying glacier change. *Global and Planetary Change* 56: 101-110.

Reist, J.D., Wrona, F.J., Prowse, T.D., Power, M., Dempson, J.B., King, J.R. and Beamish, R.J. 2006. An overview of effects of climate change on selected Arctic freshwater and anadromous fishes. *Ambio* 35, 381-387.

Reynolds, C.S. 1997. *Vegetation Processes in the Pelagic: a Model for Ecosystem Theory.* Excellence in Ecology, Book 9, ed. O. Kinne. International Ecology Institute, Oldendorf/Luhe, Germany. 371 pp.

Sala, O.E., Chapin III, F.S., Armesto, J.J., Berlow, E., Bloomfield, J., Dirzo, R., Huber-Sanwald, E., Huenneke, L.F., Jackson, R.B. Kinzig, A., Leemans, R., Lodge, D.M., Mooney, H.A., Oesterheld M., Poff, N.L., Sykes, M.T., Walker, B.H., Walker, M. and Wall D.H. 2000. Global biodiversity scenarios for the year 2100, *Science* 287: 1770-1774.

Scott, M., Gupta, S., Jáuregui, E., Nwafor, J., Satterthwaite, D. Wanasinghe, Y.A.D.S., Wilbanks, T., Yoshino, M., Kelkar, U. Mortsch L., Skea, J. 2001. Human settlements, energy, and industry. Climate change 2001: impacts, adaptation, and vulnerability. In: McCarthy, J.J., Canziani, O.F., Leary, N.A. Dokken, D.J., White, K.S. (eds) Contribution of working group I to the third assessment report of the intergovernmental panel on climate change. Cambridge University Press, Cambridge 381–416.

Shankhwar, A.K. 2010. Seasonal Variation in Physico-chemical and Biological Characteristics of Nainital Lake Water, *M.Sc. Thesis* G.B. Pant University Agriculture and Technology, 124.

Shankhwar, A.K. and Srivastava, R.K. 2011. Dioxins and Furans Formation and its degradation for safer environment, National Conference on Clean Technologies and Green Business in 21 century (CTGB-2011), 70.

Smith R.I.L. 1990. Signy Island as a paradigm of biological and environmental change in Antarctic terrestrial ecosystems. In: Kerry K.R., Hempel G. (eds) *Antarctic Ecosystems, Ecological Change and Conservation.*, Springer-Verlag, Berlin, 30–48.

Smith, R.J.L. 2001. Plant colonisation response to climate change in the Antarctic. *Geographia* 25: 19-33.

Somme, L. 1995. Invertebrates in Hot and Cold Arid Environments. Springer-Verlag, Berlin.

Straile, D. 2000. Meteorological forcing of plankton dynamics in a large and deep continental European lake. *Oecologia* 122: 44-50.

Turner J., Colwell, S.R. and Harangozo, S. 1997. Variability of precipitation over the coastal western Antarctic Peninsula from synoptic observations. *J. Geophys. Res.* 102: 13999 –14007.

Ugnar, S. 1999. Is strange weather in the air?, A study of U.S. national network news coverage of extreme weather events, *Climate Change*, 41(2): 133-150.

Upadhyay, D.S. 1995. Cold Climate Hydrometeorology, New Age International (P) Ltd, New Delhi, 345.

van Donk, E. and Kilham, S.S. 1990. Temperature effects on silicon-limited and phosphorus-limited growth and competitive interactions among three diatoms. *Journal of Phycology* 26, 40-50.

Vaughan, D.G., Marshall, G.J., Connolley, W.C., King, J.C. and Mulvaney, R. 2001. Devil in the detail. *Science* 293: 1777–1779.

Vitousek, P.M., Mooney, H.A., Lubchenco, J. and Melillo, J.M. 1997. Human domination of Earth's ecosystems. *Science* 277: 494-499.

Wah Y L. 2007. From waste to resource–Singaporés experience in NE water. In: Huber H H, Wilderer P A, Paris S, eds. *Water Supply and Sanitation for All.* Hans Huber AG Berching, 139–154.

Walther, G.-R., Post, E., Convey, P., Menzel, A., Parmesan, C., Beebee, T.J.C., Fromentin, J.-M., Hoegh-Guldberg, O. and Barlein, F. 2002. Ecological responses to recent climate change. *Nature* 416: 389-395.

Weitzman, M.L. 2007. A Review of The Stern Review on the Economics of Climate Change. *Journal of Economic Literature* 45: 703–724.

Wilbanks, T.J., Romero Lankao, P., Bao, M., Berkhout, F., Cairncross, S., Ceron, J-P, Kapshe, M., Muir-Wood, R., Zapata-Marti, R. 2007. Industry, settlement and society. Climate change 2007: impacts, adaptation and vulnerability. In: Parry ML, Canziani OF, Palutikof JP, van der Linden PJ, Hanson CE (eds) *Contribution of Working Group II to the Fourth Assessment Report of The Intergovernmental Panel on Climate Change*. Cambridge University Press, Cambridge, 357–390.

Wilderer, P.A. 2004. Applying sustainable water management concepts in rural and urban areas: Some thoughts about reasons, means and needs. *Water Science and Technology*, 49(7): 7–16.

Wilderer, P.A. 2009. Climate change and its consequences for water and wastewater management, *Front. Environ. Sci. Engin.* China 3(1): 2–11.

Winder, M. and Schindler, D.E. 2004. Climate change uncouples trophic interactions in an aquatic ecosystem. *Ecology* 85: 2100-2106.

Chapter 12

Dissolved Load Characteristics of Melt Water of Chhota Shigri Glacier, Lahaul-Spiti Valley, Himachal Pradesh, India

Virendra Bahadur Singh, A.L. Ramanathan P.G. Jose, Paramanand Sharma and Anurag Linda

Chhota Shigri glacier melt water was collected in the July – August 2008 near the snout of the glacier and samples were analysed for major cations and anions. The present study investigates chemical characteristics of melt water of Chhota Shigri Glacier in terms of association of different chemical compounds with the geology of the area. In the melt water the presence of cations varied as $c(Ca^{2+})>c(Mg^{2+})>c(Na^{+})>c(K^{+})$, while the order of concentration of anions has been $c(HCO_3)>(SO4^{2-})>c(Cl^{-})>(NO_3)$. The average pH and EC of melt water of Chhota Shigri Glacier was recorded as 7.3 (6.7-7.9) and 20.6µs/cm (14-27 µs/cm) respectively. The average concentration range of major cations was found to be Ca^{2+} 80.5µeq/l (68.3-95.5µeq/l) and Mg^{2+} 67.1µeq/l (46.1-88.1 µeq/l) followed by Na^{+} 20.1µeq/l(13.5-28.5 µeq/l)and K^{+} 15.8 µeq/l (11.1-21.6 µeq/l) respectively. The average concentration of major anions was HCO_3^- 125.9 µeq/l (60.7-184.6 µeq/l) followed by SO_4^{2-} 54.4 µeq/l (25.0-74.8 µeq/l), Cl^{-}14.1 µeq/l (6.1-22.7 µeq/l) and NO_3 1.5 µeq/l (0.00-8.3 µeq/l) respectively. The average TSM of Chhota Shigri Glacier melt waters was 597.4

mg/l (149-1310 mg/l) and dissolved silica was 26.1μm/l. The charge balance between anions and cations of the samples averaged -3.0. The relative high contribution of $c(Ca^{2+}+Mg^{2+})$/ Total cations (TZ^+) and high $c(Ca^{2+}+Mg^{2+})/c(Na^++K^+)$ ratio (4.1) and low $c(Na^++K^+)$/Total cations(TZ^+) ratio (0.2), indicate that the melt water chemistry of the Chhota Shigri Glacier is mostly controlled by carbonate weathering. The estimated bicarbonate contribution from carbonate and silicate weathering using the equation suggested by Raymahasay (1986) indicate that most of the HCO_3 (68.7 per cent) comes from carbonate weathering and remaining(31.3 per cent) from silicate weathering.

Introduction

The Himalayas are the largest storehouse of snow and ice outside the polar region and are commonly referred to as "the abode of eternal snow" (Vohra, 1993). The Himalayas, with an average elevation of 6000 m are the repositories of some of the highest and largest glaciers of the world. It has been estimated that about 38221 km^2 of the Himalayan ranges are glaciated (Bahadur, 1988). There are more than 5222 glaciers in the Himalaya, scattering in three river systems, *i.e.* Indus, Ganga and Brahmaputra (Puri, 1994). The melt water discharge from Himalayan glaciers is the major source of water for nearly half a billion population of the country. This area is considered geo-dynamically active and it is prone to violent crustal movements causing high erosion rates and seismicity (Valdiya, 1998). Various studies of glaciers specially in Himalaya by Ahmad and Hasnain (1999), Kumar *et al.* (2008) and in Alpine region by Rainwater and Guy (1961), Beherens *et al.* (1971), Lorrain and Souchez (1972), Collins (1978), Tranter and Raiswell (1991), have suggested that principal flow component in water passing through subglacial channels had good exposure to subglacial environment. These studies indicate high rates of chemical weathering in glaciated catchments resulting from rapid dissolution of freshly ground glacial rock flour. The chemical enrichment of melt water released from the terminus of glacier is usually explained in terms of acquisition and transport of solutes during passage across subglacial surfaces (Collins 1981). The acquisition of solutes includes the chemical weathering of rocks occurring in the glacial environment whereby the water bodies gets enriched with several cationic and anionic constituents such as Ca^{2+}, Mg^{2+}, Na^{+}, K^{+}, HCO^{-}_{3} and SO_4^{2-} (Rainwater and Guy, 1961; Collins, 1978; Hasnain and Thayyen, 1996). Ions from these primary sources precipitate out of solution to form secondary sources, which include subglacial precipitates such as carbonates (Hallet, 1976). However, the main source of ions in glacial melt water is weathering of rock forming minerals, with little contribution of atmospheric sea salt and atmospheric inputs (Berner and Berner, 1987). When water flows at the ice-rock interface, weathering of the bedrock is likely to occur.

The morphology, bedrock topography, meteorological parameters, hydrogeochemical as well as the dynamics of Chhota Shigri glacier have been studied by Dobhal *et al.* (1995), Kumar and Dobhal (1997), Kumar (1999), Kulandaivelu *et al.* (1989), Hasnain *et al.* (1989). Chandra river of Himachal Pradesh is one of the important sources of water fed by several glaciers including Chhota Shigri glacier. Information available on mass balance, discharge and chemistry of snow, ice and melt water of Chhota Shigri glacier is sparse and inadequate and needs to be substantiated by comprehensive studies.

Study Area

Geographically Chhota-Shigri glacier is located between 32° 11′-32° 17′N and 77° 29′- 77° 33′E, it lies on the Chandra river basin on the northern ridge of Pir Panjal range in the Lahaul–Spiti valley of Himachal Pradesh, India. Table 12.1 gives list of geographical and tropographical characteristics of Chhota Shigri glacier. Chhota Shigri glacier is fed by four tributaries one coming from east side of accumulation area and flowing on the right bank of glacier and the second coming from the west side and flowing on the left bank of the glacier originating in the vicinity of the 5500 to 6000m a.m.s.l. peak. The ablation area is totally debris-covered below 4350m a.m.s.l. Figure 12.1 gives a map showing the location of study area.

The ablation zone of Chhota Shigri glacier is longitudinally divided into two by a prominent superficial medial morain. The western flank has ice flow direction W-E in upper ablation zone, while the eastern flank has E-W flow direction. In the lower ablation zone, the ice flow on both the flanks changes to a SSW-NNE direction.

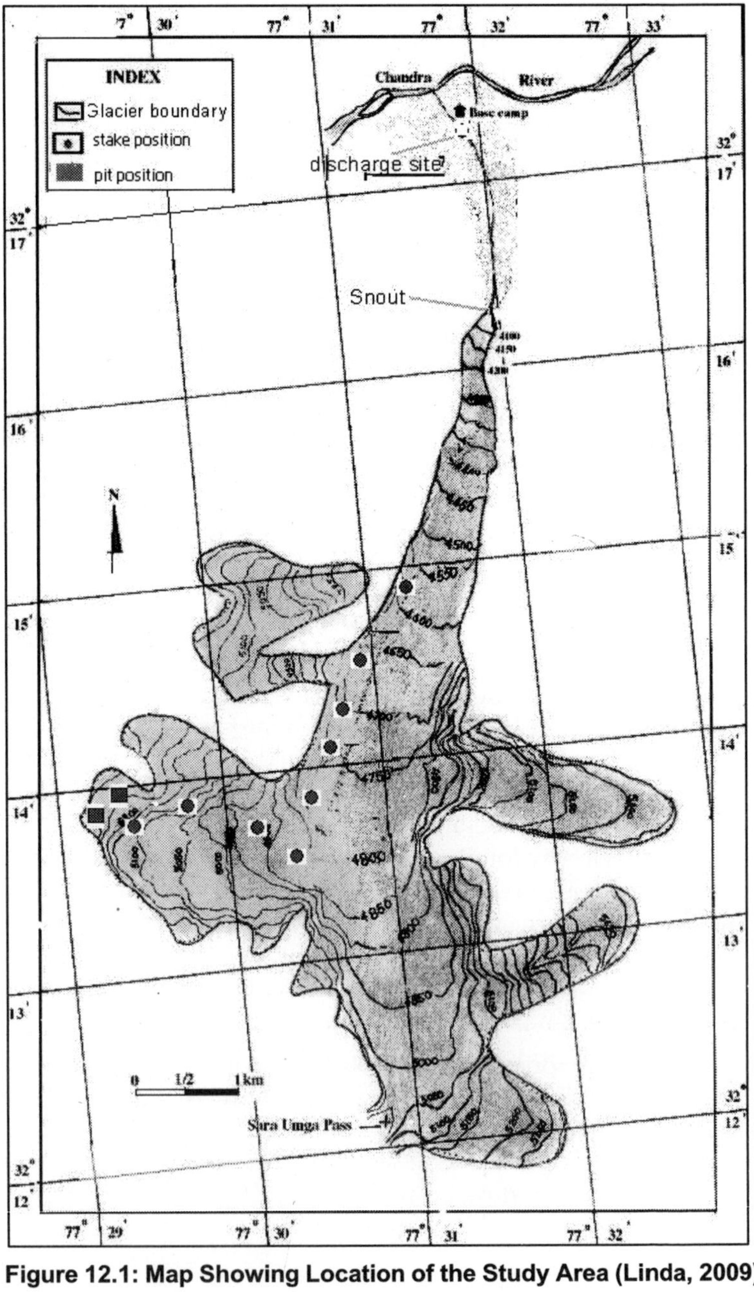

Figure 12.1: Map Showing Location of the Study Area (Linda, 2009)

Table 12.1: List of Geographical and Topographical Characteristics of Chhota Shigri Glacier

Latitude	32º 11′-32º 17′N
Longitude	77º 29′- 77º 33′E
Max. elevation	6263m a.m.s.l
Snout position	4050m a.m.s.l
Basin area	34.70km² (47 per cent glacierized) at 3900m a.m.s.l on the proglacial stream.
Total glacierized area	16.4 km²
Chhota Shigri glacier area	15.7 km²
Glacier length	9 km
Mean orientation	North

Source: (Wagnon *et al.*, 2007)

Geology of Area

Chhota Shigri glacier lies within the central crystallines of the Pir Panjal range of the Himachal Himalaya. Here meso- to keta-zonal metamorphites, migmatites and gneisses (Kumar *et al.*, 1997) are found. In few places, granite rocks of different composition and younger age indicate rejuvenation but 3 km upstream of Chhota Dara, in upper Chandra valley, older Paleozoic granitic rocks are exposed. The Haimanta formation overlies these within a tectonic break, where black slates, phyllites and fine grained biotite schists are exposed (Kumar *et al.*, 1997). The slates and phyllites shows a well developed thrust tectonic contact, which forms the crest of the north ridge. Box type flods with decollement are quite prominent in the Haimanta formation. The Haimantas, which rest directly on basement rock, are highly metamorphosed metasediments and show intense folding and shearing. The brown biotite, with a fine grained texture, shows intense heating effect, which indicates periodic re-heating of the granite rock below.this is also indicated by the various types of granite and gneiss rocks present in the basement. Schistose gneiss and augen gneiss has developed in the granite without any distinct margins. Chhota Shigri glacier rests on the granitic basement rocks.

Materials and Methods

Meltwater samples were collected from Chhota Shigri glacier during July, August. 2008. Electrical conductivity (EC) was measured by using Pentex EC meter. pH of all samples were measured by Consort microcomputer(P-307) ion meter. Suspended sediments were separated from the water samples in the laboratory by using 0.45 micro millipore membrane filters of 47mm diameter. Vacuum pump was used for accelerate the filtration. Bicarbonate was determined by following potentiometric titration method. (APHA, 1985), Chloride ion concentration was measured by the mercury (II) thiocynate method, Sulphate concentration was measured by turbidimetric method (APHA, 1985), The dissolved silica was determined by molybdosilicate method (APHA, 1985), Nitrate concentration in water samples was determined by brucine-sulphanilic acid method and The content of major cations Na^+, K^+, Ca^2, Mg^{2+} was determined on GBC 906 atomic spectrophotometer (AAS).

Results and Discussion

The chemical composition of melt water of Chhota Shigri glacier is shown in Table 12.2. This table also shows TDS/EC ratio (0.7). The charge balance is calculated by the formula (TZ^+-TZ^-/TZ^++TZ^-) X100 between cations and anions, which is <10 per cent are within acceptable limits and confirming the reliability of the analytical results.

The average value of EC and pH of melt water Chhota Shigri glacier was 20.6μs/cm (range: 14-27μs/cm) and 7.3 (range: 6.7-7.9) respectively. The EC of meltwater in the catchments of glacier underlain by igneous and metamorphic rocks is generally 6-60 μs/cm (Mmeybeck 1984). The pH was observed to be mildly alkaline in nature. The pH of melt water is 7.2 for Kafni glacier (Singh, A.K. *et al.*, 1998). The pH of Himalayan glacier meltwaters is significantly higher than pH of the melt water of the North Cascade glaciers (Reynold and Johnson, 1972).

The concentration of major cations are recorded as Ca^{2+} 80.5 μeq/l (68.3-95.5 μeq/l), Mg^{2+} 67.1 μeq/l (46.1-88.1 μeq/l), Na^+ 20.1 μeq/l (13.5-28.5 μeq/l) and K^+ 15.8 μeq/l (11.1-21.6 μeq/l). The average chemical composition of melt water showed major percentage of Ca^{+2} (43.8 per cent of total cations) followed by Mg^{2+} (36.5 per cent of total cations), Na^+ (10.9 per cent of total cations) and

Table 12.2: Chemical Characteristics of Melt Water of Chhota Shigri Glacier

Sl. No.	Description	EC	PH	TSM	HCO_3	Cl	SO_4	NO_3	H_4SiO_4	Na	K	Ca	Mg	TZ+	TZ-	TDS	TDS/EC	Balance
1.	2008/7/30/7.00/DS	21	7.3	390.0	138.0	16.6	52.0	0.0	30.6	19.2	13	70.1	68.3	170.6	206.6	14.7	0.70	-9.5
2.	2008/7/30/12.00/DS	21	7.1	427.0	153.0	18.0	29.9	3.3	31.7	16.5	14	77.4	60.9	168.8	204.2	14.8	0.71	-9.5
3.	2008/7/30/17.00/DS	21	7.3	637.0	142.7	13.1	38.3	0.0	33	16.7	13.3	76.2	67.8	174.0	194.1	14.2	0.68	-5.5
4.	2008/7/31/7.00/DS	22	7.2	305.5	138.0	14.7	46.0	2.0	15.2	20.4	14.3	90.3	83	208.0	200.7	15.1	0.69	1.8
5.	2008/7/31/12.00/DS	21	7.4	1120.0	122.0	19.9	54.3	3.8	20.5	20.1	15.3	90	62.6	188.0	200.0	14.6	0.70	-3.1
6.	2008/7/31/17.00/DS	20	7.0	153.0	92.0	22.2	66.9	0.0	21	22.5	20.7	82.4	75.6	201.2	181.1	13.5	0.67	5.3
7.	2008/7/1/7.00/DS	19	7.6	281.0	105.5	16.7	52.1	0.0	35.8	23.4	21.6	85.4	81.2	211.6	174.3	13.6	0.72	9.7
8.	2008/7/1/12.00/DS	15	7.9	533.0	65.9	8.1	65.3	0.9	15.4	25.8	16.4	77.1	50.4	169.7	140.2	10.9	0.73	9.5
9.	2008/7/1/17.00/DS	19	7.1	1113.0	117.9	16.8	50.3	2.1	17.3	16.2	11.9	81.9	58.3	168.3	187.1	13.5	0.71	-5.3
10.	2008/8/2/7.00/DS	26	7.1	260.0	182.4	6.2	68.0	0.0	36	21	16.1	93.0	87.0	217.1	256.6	18.6	0.72	-8.3
11.	2008/8/2/12.00/DS	24	7.4	264.5	143.0	12.8	66.1	8.2	16.9	17.5	18.8	79.1	72.1	187.5	230.1	16.4	0.69	-10.2
12.	2008/8/2/17.00/DS	18	6.7	149.0	89.0	18.0	62.5	0.1	18.8	20.1	17.3	76.1	65.4	178.9	169.6	12.5	0.70	2.7
13.	2008/8/3/7.00/DS	22	7.4	307.6	140.0	16.9	48.0	3.5	17.3	21.9	15.5	95.4	84.2	217.0	208.4	15.7	0.71	2.0
14.	2008/8/3/12.00/DS	22	7.3	1237.0	150.0	6.6	59.4	0.0	37.9	16.1	15.9	88.2	62.5	182.7	216.0	15.7	0.72	-8.4
15.	2008/8/3/17.00/DS	20	7.2	1118.0	120.0	17.8	52.5	2.4	18.4	18.6	14.2	85.1	61.5	179.4	192.7	14.0	0.70	-3.6

Contd...

Table 12.2–*Contd...*

Sl. No.	*Description*	*EC*	*PH*	*TSM*	*HCO_3*	*Cl*	*SO_4*	*NO_3*	*H_4SiO_4*	*Na*	*K*	*Ca*	*Mg*	*TZ+*	*TZ-*	*TDS*	*TDS/EC*	*Bala-nce*
16.	2008/8/4/7.00/DS	27	7.1	256.0	184.6	7.6	70.7	0.0	38.0	22.2	18.7	95.5	88.1	224.5	262.9	19.1	0.71	-7.9
17.	2008/8/4/12.00/DS	22	7.3	1310.0	130.5	18.3	60.3	2.4	40.0	19.8	16.2	75.5	65.2	176.7	211.4	15.0	0.68	-8.9
18.	2008/8/4/17.00/DS	20	7.2	1141.0	130.5	10.8	48.3	1.5	16.7	18.7	12.4	79.5	48.4	158.9	191.2	13.8	0.69	-9.2
19.	2008/8/5/7.00/DS	23	7.5	268.5	140.9	11.7	63.5	8.3	15.0	16.5	16.3	81.6	69.3	183.7	224.4	16.1	0.70	-10.0
20.	2008/8/5/12.00/DS	20	7.3	451.2	150.0	11.2	25.0	0.0	32.4	17.7	14.0	69.3	64.6	165.5	186.2	13.9	0.69	-5.9
21.	2008/8/5/17.00/DS	21	7.0	425.0	150.0	16.0	27.8	5.2	31.6	15.5	12.8	75.2	58.9	162.5	199.0	14.4	0.69	-10.1
22.	2008/8/6/7.00/DS	23	7.5	154.0	140.0	10.7	69.4	0.0	34.9	25.4	16.6	89.7	75.5	207.2	220.1	16.2	0.70	-3.0
23.	2008/8/6/12.00/DS	21	7.4	639.0	145.7	15.3	40.5	0.0	33.3	17.8	14.5	78.5	69.9	180.7	201.4	14.8	0.70	-5.4
24.	2008/8/6/17.00/DS	20	6.8	582.0	120.0	22.7	44.9	0.0	30.1	15.6	18.6	70.5	55.0	159.6	187.6	13.4	0.67	-8.1
25.	2008/8/7/7.00/DS	25	7.6	183.0	150.0	21.7	70.6	0.0	34.3	25.8	16.8	90.8	80.6	213.9	242.3	17.3	0.69	-6.2
26.	2008/8/7/12.00/DS	21	7.2	1300.0	141.0	8.2	52.1	0.0	32.9	17.8	14.0	69.3	58.4	159.5	201.3	14.4	0.69	-11.6
27.	2008/8/7/17.00/DS	19	7.2	1120.0	120.0	14.6	50.8	0.0	31.3	13.5	16.6	70.0	52.8	152.8	185.4	13.3	0.70	-9.6
28.	2008/8/8/7.00/DS	22	7.5	280.0	149.6	14.6	49.9	0.0	34.7	22.5	20.5	85.3	76.1	204.4	214.1	16.0	0.73	-2.3
29.	2008/8/8/12.00/DS	22	7.5	393.0	140.0	18.8	58.1	0.0	31.7	20.7	15.2	68.3	73.5	177.6	216.9	15.3	0.70	-9.9
30.	2008/8/8/17.00/DS	18	6.7	430.0	73.4	15.3	74.8	0.0	16.1	22.7	14.9	80.9	80.4	198.9	163.6	12.3	0.68	9.7

Contd...

Table 12.2–*Contd...*

Sl. No.	Description	EC	PH	TSM	HCO_3	Cl	SO_4	NO_3	H_4SiO_4	Na	K	Ca	Mg	TZ+	TZ-	TDS	TDS/EC	Balance
31.	2008/8/9/7.00/DS	19	6.8	150.0	90.0	20.1	64.8	0.0	20.0	21.2	18.5	79.2	70.6	189.4	174.9	13.0	0.68	4.0
32.	2008/8/9/12.00/DS	15	7.8	528	63.9	7.1	63.2	0.8	14.3	28.5	14.3	70.0	55.2	168.0	134.9	10.5	0.70	10.9
33.	2008/8/9/17.00/DS	19	7.0	1136.0	128.5	9.3	46.0	1.2	14.0	17.7	11.1	75.2	46.1	150.1	185.0	13.3	0.70	-10.4
34.	2008/8/10/7.00/DS	17	6.9	435.0	75.4	16.8	61.1	0.3	18.8	21.6	16.1	80.1	72.5	190.3	153.6	11.7	0.69	10.7
35.	2008/8/10/12.00/DS	14	7.6	525.0	60.7	6.1	61.1	0.5	12.1	27.2	13.2	70.0	48.1	158.5	128.4	10.0	0.71	10.5
36.	2008/8/10/17.00/DS	24	7.5	1240.0	153.1	7.6	63.5	0.3	40.4	17.2	17.0	95.2	64.5	193.9	224.5	16.4	0.68	-7.3
	MAX	27	7.9	1310.0	184.6	22.7	74.8	8.3	40.4	28.5	21.6	95.5	88.1	224.5	262.9	19.1	0.73	10.9
	MIN	14	6.7	149.0	60.7	6.1	25.0	0.0	12.1	13.5	11.1	68.3	46.1	150.1	128.4	10.0	0.67	-11.6
	AVER.	20.6	7.3	597.4	125.9	14.1	54.7	1.5	26.1	20.1	15.8	80.5	67.1	183.5	196.4	14.4	0.70	-3.0

Unit: Dissolved ions in μeq/l, TZ^+,TZ^- in μeq/l, H_4SiO_4 in μmole/l and TDS and TSM in mg/l.

K^+ (8.6 per cent of total cations). The concentration of major anions are HCO_3^- 125.9µeq/l (60.7-184.6 µeq/l), SO_4^{2-} 54.7µeq/l (25.0-74.8 µeq/l), Cl^- 14.1µeq/l (6.1-22.7 µeq/l) and NO_3^- 1.5µeq/l (0.00-8.3 µeq/l). The average chemical composition of melt water showed major percentage of HCO_3^- (64.2 per cent of total anions) followed by SO_4^{2-} (27.8 per cent of total anions), Cl^- (7.2 per cent of total anions) and NO_3 (0.8 per cent of total anions). The melt water of Chhota Shigri glacier was found to contain 12.1-40.4 µmole/l of dissolved silica with average value 26.1 µmole/l. Total suspended matter (TSM) was found to range 149.0-1310.0 mg/l with an average value 597.4 mg/l. The major cations contribution was found to be nearly 48.3 per cent of the total dissolved solid (TDS) while major anion contribution was nearly 51.7 per cent of the total dissolved solids (TDS). TDS content is generally low. The main reason for the low solute content is the presence of granite rocks below the glacier, as they are less prone to chemical weathering.

Source of Dissolved Ions

The chemical composition of glacier runoff indicates the nature of chemical weathering taking place beneath the glacier (Lorrains and Souchez, 1972; Reynolds and Johnson, 1974; Raiswell, 1984). Catchments of glaciers undergo more intense chemical weathering than the catchments which do not have glaciers (Reynolds and Johnson, 1972 and Collins, 1983). The solute content present in melt water is derived from two sources, lithospheric and atmospheric. The release of solute to glacier melt waters is governed by the interaction between the rock water interface, types of weathering reactions which is taking place, mineralogy of the bed rock, flow rate of water and the composition of atmospheric deposition in the catchment of glacier.

The chemical composition of melt water with reference to major ions can be explained on the basis of weathering of various rock forming minerals. In the glacier environment, the proton H^+ used for chemical weathering of carbonates, sulphides and aluminosilicates are mainly supplied from the equilibration of melt water with atmospheric CO_2 and/or by oxidative weathering of sulphides (Raiswell, 1984). Hence Ca^{+2}, Mg^{+2} and HCO_3^- in aquatic environment can be derived from the weathering of carbonate and/or silicate by the reaction.

$CO_2 + H_20 \rightarrow H_2CO_3$

$H_2CO_3 \rightarrow H^+ + HCO_3^-$

$4\,FeS_2 + 15\,O_2 + 8\,H_2O \rightarrow 2\,Fe_2O_3 + 8\,SO_4^{2-} + 16\,H^+$

Carbonate minerals notably Calcite and Dolomite reacts with carbonic acid and released HCO_3^- and cations to the solution by the reactions.

$CaCO_3 + H_2CO_3 \rightarrow Ca^{2+} + 2\,HCO_3^-$

$Ca\,Mg\,(CO_{3)} + 2\,H_2CO_3 \rightarrow Ca^{2+} + Mg^{+2} + 4\,HCO_3^-$

(Ca, Mg, Na, K) Silicates $+ H_2CO_3 \rightarrow Ca^{2+} + Mg^{+2} + Na^+ + K^+ + H_4SiO_4 + HCO_3^- +$ Clay minerals

Hence two major anions HCO_3^- and SO_4^{2-} in melt water of glacier are mainly derived from the dissolution of atmospheric CO_2 in H_2O and the oxidation of sulphides or dissolution of gypsum. The dissolution of carbonate rocks proceeds more rapidly than silicate breakdown and is the likely mechanism of solute acquisition. The relative proportion of the various ions in solution is depending on their relative abundance in the host rock and their solubility. The Ca+Mg Vs TZ^+ plot in Figure 12.2(a) shows that the concentration of $Ca^{2+} + Mg^{+2}$ mostly fall along 1:1 line with an average equivalent ratio of 0.8. It is indicated that the carbonate weathering could be the major source of dissolved ions in the melt water of Chhota Shigri glacier. The $Na^+ + K^+$ Vs TZ^+ plot in Figure 12.2(b) shows that values mostly fall below 1:1 line with low equivalent ratio 0.2, which indicates that there is a small contribution from silicate weathering. The relative high contribution of Ca+Mg to the total cations, the high Ca+Mg/Na+K ratio(4.1) and the low $(Na+K)/TZ^+$ ratio (0.2) also confirms that carbonate weathering was the major source of dissolved ions in the melt water of Chhota Shigri glacier. The estimation of carbonate and silicate weathering done by using the Raymahasay (1986) method indicates that 68.7 per cent of bicarbonate ions come from carbonate weathering and 31.3 per cent from silicate weathering.

A plot of Ca+Mg Vs HCO_3 in Figure 12.3 (a) shows that in most of the samples the Ca+Mg content is slightly in excess of HCO_3 suggesting that excess of Ca+Mg in the meltwater of glacier should be balanced by Cl-$^-$ and SO_{4-}^{2-}. Among alkali metals, K^+ in the melt water of glacier is derived from the lithogenic input, where as Na^+

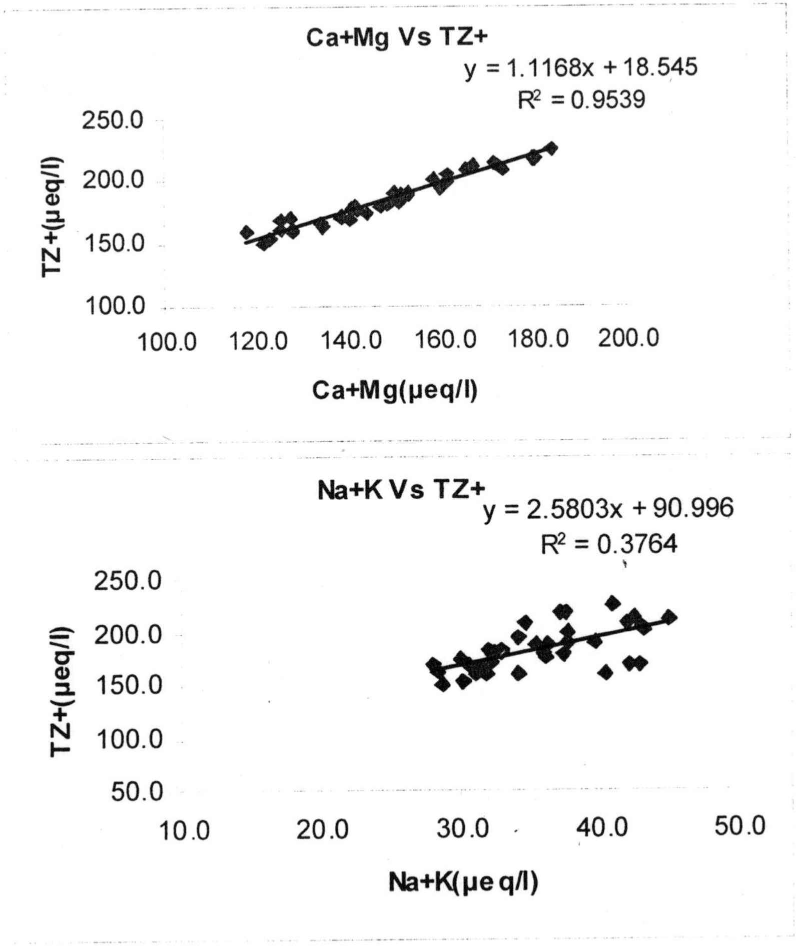

Figure 12.2 (a) Scatter Diagram Between Ca^{2+}+Mg^{2+} Vs Sum of Cations (TZ^+) (b) Scatter Diagram Between Na^++K^+ Vs Sum of cations (TZ^+)

may be of lithogenic or atmospheric origin. Low concentration of NO_3^- at the sampling site was reported, which indicated that nitrate is derived exclusively from the atmosphere associated with the acidic nitrate aerosols.

However, the low concentration of Cl^- and excess of Na+K over Cl^- in Figure 12.(3b) and (Na+K)/Cl ratio 2.5 indicate that in the study area cyclic contribution is not important and most of the Na^+

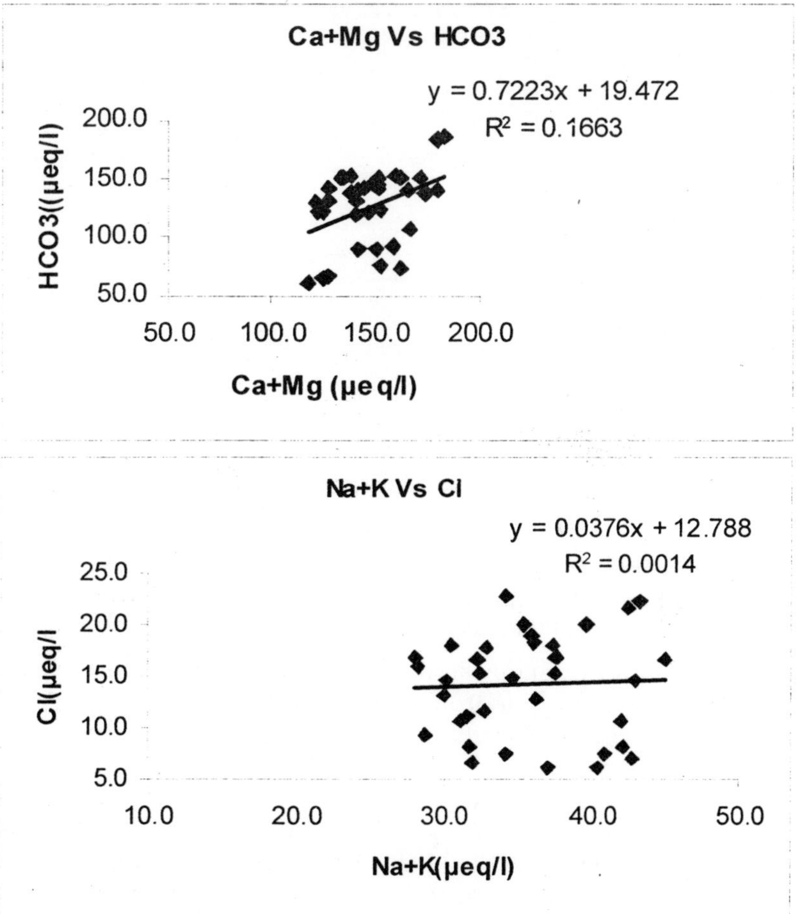

Figure 12.3: Scatter Diagram Between (a) Ca^{2+}+Mg^{2+} Vs HCO_3 (b) Na^++K^+ Vs Cl^-

and K^+ could be derived from slilicate weathering. The observed Na/Cl and K/Cl ratio may also be indicative of contribution from salt dissolution such as evaporates. Low concentration of NO_3^- at some of the sampling points can be attributed to little anthropogenic activities.

In the Gibbs plot (Figure 12.4) most of the samples fall outside the boomerang region reflecting the low TDS of melt water of glacier, thus the Gibbs (1970) assumption that low to moderate levels of TDS

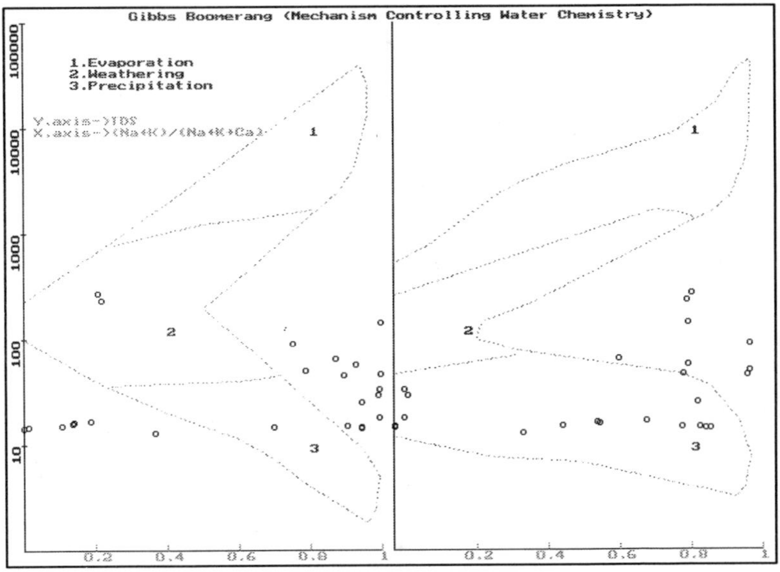

Figure 12.4: Variation of Weight Ratio (Na^+/Na^++Ca^{2+}) as a Function of TDS (after Gibbes 1970)

and sodium are primarily derived by precipitation may not be valid, specially in glacier environment, since there can be significant weathering product of silicate rocks under certain circumstances (Singh and Hasnain, 1998). The thermodynamic stability relationships of water chemistry in the Na^+, K^+, Ca^{2+} and Mg^{2+} silicate system is demonstrated in Figure 12.5 and Figure 12.6 (after Garrels and Christ, 1965). The pH-lgc(H_4SiO_4) relationship demonstrates that chemical composition of the meltwater of glacier is in kaoline region, which indicates that silicate weathering also is significant in the glacier basin.

A parametric correlation analysis for average values of all measured parameters is given in Table 3. HCO_3^- is highly correlated with EC ($r^2 = 0.90$), TZ^-($r^2 = 0.92$) and TDS ($r^2 = 0.91$), Ca^{2+} with TZ^+(r^2 =.83) and Mg^{2+} with TZ^+($r^2 = 0.92$).

The relationship between rock types and water composition can be evaluated by the trilinear Piper Diagram (Piper 1944). The piper diagram is an ingenious construction, which consists of two triangular diagrams at the lower left and lower right, describing the relative composition of cations and anions and intervening diamond

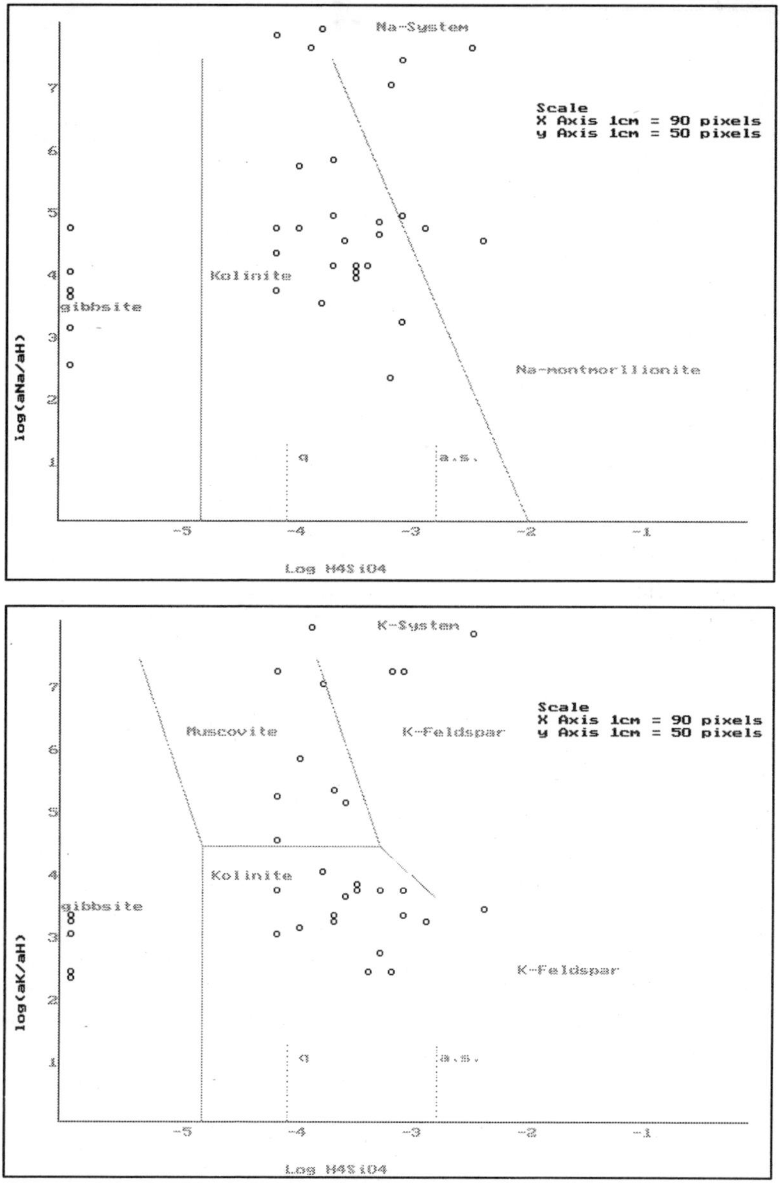

Figure 12.5: Stability Diagram for Na-K System

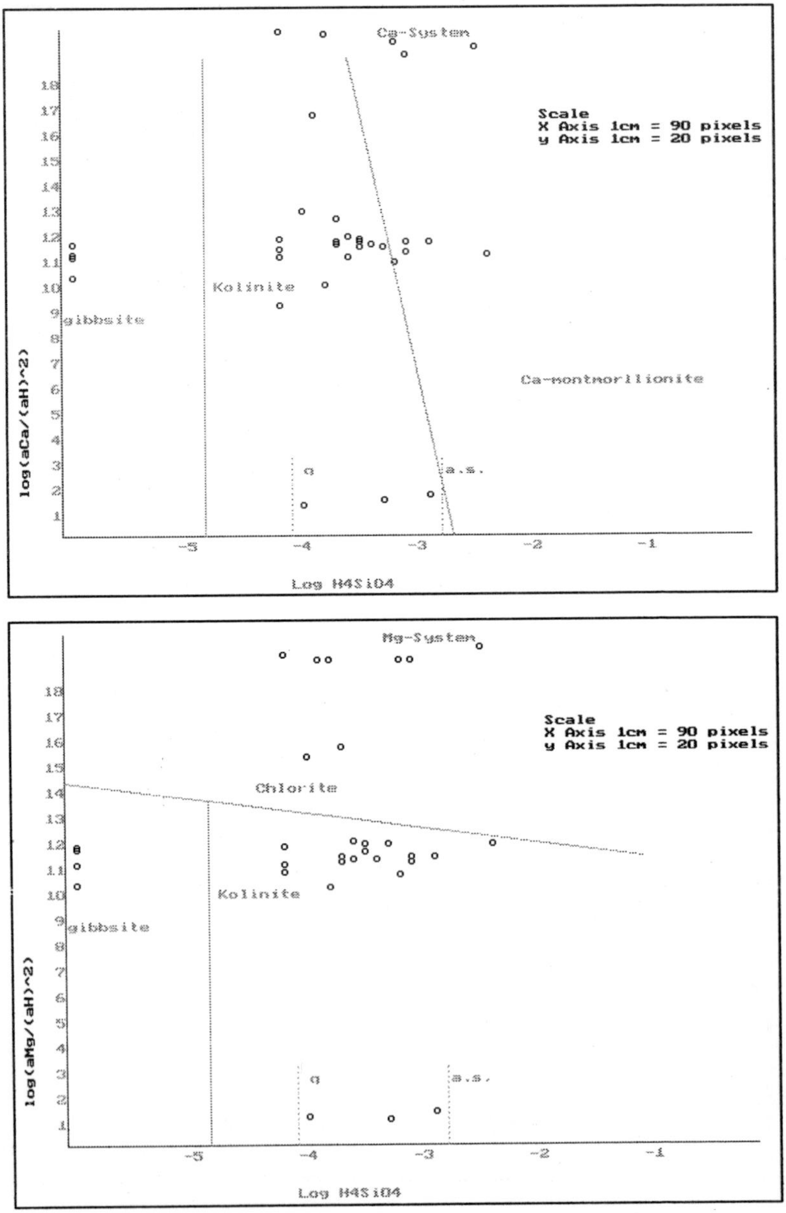

Figure 12.6: Stability Diagram for Ca-Mg System

Table 12.3: Parametric Correlation Matrix of Measured Parameters of Chhota Shigri Glacier Meltwater

	EC	*pH*	*TSM*	*HCO_3*	*Cl*	*SO_4*	*NO_3*	*H_4SiO_4*	*Na*	*K*	*Ca*	*Mg*	*TZ+*	*TZ-*	*TDS*
EC	1														
pH	-0.01	1.00													
TSM	-0.10	0.02	1.00												
HCO_3	0.90	0.02	0.06	1.00											
Cl	0.00	-0.36	-0.22	-0.12	1.00										
SO_4	0.09	0.10	-0.19	-0.28	-0.14	1.00									
NO_3	0.19	0.11	-0.03	0.16	0.06	-0.11	1.00								
H_4SiO_4	0.58	0.09	0.10	0.64	-0.02	-0.14	-0.36	1.00							
N	-0.27	0.42	-0.45	-0.46	-0.13	0.55	-0.29	-0.23	1.00						
K	0.23	0.04	-0.44	-0.03	0.26	0.43	-0.10	0.28	0.24	1.00					
Ca	0.58	0.07	-0.13	0.38	-0.06	0.35	0.08	0.15	0.18	0.32	1.00				
Mg	0.60	-0.06	-0.61	0.36	0.20	0.33	-0.06	0.29	0.26	0.50	0.62	1.00			
TZ+	0.55	0.07	-0.53	0.27	0.10	0.48	-0.06	0.22	0.43	0.58	##	**0.92**	1.00		
TZ-	0.99	0.01	-0.06	**0.92**	-0.02	0.08	0.20	0.58	-0.30	0.18	0.53	0.53	0.49	1.00	
TDS	0.99	0.04	-0.11	**0.91**	-0.05	0.10	0.17	0.59	-0.23	0.24	0.62	0.61	0.58	**0.99**	1

shaped diagram that combines the composition of cations and anions. The Piper Plots (Figure 12.7) tells, most of the samples fall towards the Ca^{2+} and Mg^{2+}apex and HCO_3 apex. This further confirms that carbonate weathering of Calcite, Dolomite, etc. from freshly exposed rocks due to rapid erosion by glacier dominates over silicate weathering as indicated above.

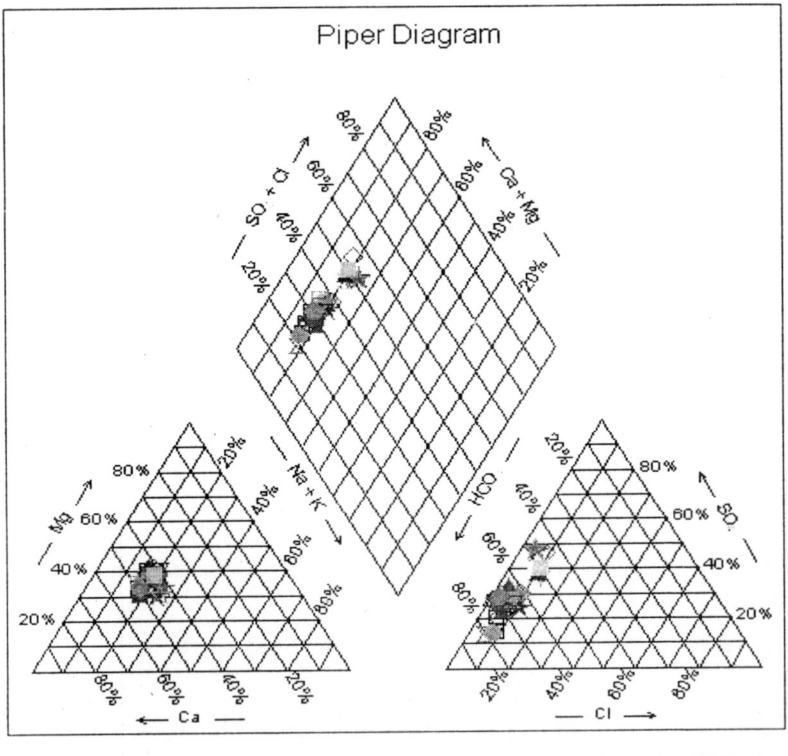

Figure 12.7: Piper Plot of Dissolved ion Concentrations for Chhota Shigri Glacier Meltwater

Conclusion

Meltwater of Chhota Shigri glacier is slightly alkaline in nature. HCO_3 is dominant anion and Ca^{2+} and Mg^{2+} are dominant cations. High (Ca + Mg/Na + K) and low (Na + K/TZ^+) ratio indicate that carbonate weathering was the major source of dissolved ions with a small contribution from silicate weathering. TDS content is generally low; the main reason for low solute content is due to presence of granite rock below the glacier as they are less prone to chemical weathering. Ca^{+2}, Mg^{+2} and HCO_3^- are derived from the weathering of carbonate and/or silicates and SO_4^{2-} by the oxidation of sulphides or dissolution of gypsum. The stability diagram relationship demonstrates that chemical composition of the melt water of glacier is in the region of Kaolinite, indicating high degree of chemical weathering in this low temperature high energy environment.

References

Ahmad, S., and Hasnain, S.I. 1999. Hydro-Geochemical characteristics of meltwater draining from Bagni glacier, Dhauliganga basin, Garhwal Himalayas, Uttar Pradesh. J *Appl Hydrol XII* (2 and 3):53–60.

APHA. 2005. *Standard Methods for Examination of Water and Wastewater*, 21st edn. American Public Health Association, Washington, 6710 pp.

Bahadur J. 1988. Himalayan water from snow and glaciers. In *First National Water Conference Proceedings*, Vol. II. Ministry of water resources, Government of India: 59 – 65.

Behrens, H., Bergmann, H., Moser, H., Rauert, W., Stichler, W., Eisner, H. and Pessel, K. 1971. Study of the discharge of alpine glaciers by means of environmental isotopes and dye tracers. *Zeitscherkd Glazialgeo.* 7:79–102.

Berner, E.K. and Berner, R.A. 1987. *The Global Water Cycle: Geochemistry and Environment*. Prentice Hall, Englewood Cliffs, 397 pp.

Collins, D.N. 1978. Hydrology of an alpine glacier as indicated by the chemical composition of meltwater. *Zeitschrift fur Gletscherked Glazialgeologie.* 13:219–238.

Collins, D.N. 1981. Seasonal variation of solute concentration in meltwaters draining from an alpine glacier. *Ann Glaciol.* 1:11–16.

Dhobal. D.P., Kumar, S. and Mundepi, A.K. 1995. Morphology and glacier dynamics studies in monsoon- arid transition zone: An example from Chhota Shigri glacier, Himachal Himalaya, India. *Current Science.* 68(9):936 – 944.

Garrels, R.M. and Christ, C.L. 1965. Solution minerals and equilibrium New York, Harper and Row. 450.

Gibbs, R.J. 1970. Mechanism controlling world water chemistry. *Science*. 170:1088-1090.

Hallet, B. 1976. Deposits formed by subglacial precipitation of $CaCO_3$. *Geol Soc Am Bull* 87: 1003–1015.

Hasnain, S.I. Subramanian, V. and Dhanapal, K. 1989. Chemical characteristics of meltwaters from a Himalayan Glacier in India. *Journal of Hydrology*. 106:98-108.

Hasnain, S.I and Thayyen, R. 1996. Sediment transport and solute variation in meltwater of Dokriani glacier (Bamak), Garhwal Himalaya. *Geol Soc India*. 47:731–739.

Kumar, S. and D.P. Dobhal. 1997. Climatic effects and bedrock control on rapid fluctuations of Chhota Shigri glacier, northwest Himalaya, India. *J. Glaciol*. 43(145):467-472.

Kumar, S. 1999. Chhota Shigri Glacier: Its kinematic effects over the valley environment, in the northwest Himalaya. *Current Science*. 77(4):594-598.

Kulandaivelu, E. Ray, B. and Sharma, J.K. 1989. Meteorological report of Chhota Shigri glacier expedition, 1988, Multi disciplinary glacial expedition to Chhota Shgri. 1988. Technical report no. 3, pp. 37-57.

Kumar, K., Miral, M., Joshi, S., Pant, N., Joshi, V. and Joshi, L.M. 2008. Solute dynamics of meltwater of Gangotri glacier, Garhwal Himalaya, India. *Environ Geol*. 58:1151–1159.

Linda, Anurag. 2009. Snow and ice mass budget of the Chhota Shigri Glacier, Lahaul and Spiti, Himachal Pradesh 2003-2007; Ph.D. thesis, Jawaharlal Nehru Univeristy 187p. (unpublished).

Lorrain, R.D. and Souchez, R.A.1972. Sorption as a factor in the transport of major cations by meltwater from and Alpine glacier. *Quarter Res*. 2(2):253–256.

Meybeck, M. 1981. Pathways of major elements from land to ocean through river. In: River inputs to ocean systems. UNEP/ UNESCO report, pp. 18-30.

Puri, V.M.K. 1994. Element of glaciology. In *Geological Survey of India*. Foundation Course in Glaciology. 79.

Rainwater, F.H. and Guy, H.P. 1961. Some observations on the hydrochemistry and sedimentation of the Chamberlain Glacier

Area Alaska. *United States Geological Survey Professional Paper* 414- c, pp cl–c14.

Raiswell, R. 1984. Chemical models of solute acquisition in glacial meltwater. *J Glaciol* 30(104):49–57.

Reynolds, R.C. and Johnson, N.M. 1972. Chemical weathering in the temperate glacial environment of the Northern Cascade mountains. *Geochim Cosmochim Acta*.36:537–554.

Raymahasay, B.C. 1986. Geochemistry of bicarbonate in river water. *J. Geol. Soc. India*. 27: 114-118.

Singh, A.K. and Hasnain, S.I. 1998. Major ion chemistry and weathering control in a high altitude basin: Alaknanda river, Garhwal Himalaya, India. *Hydrol Sci*. 43(6):825–843.

Tranter M, Raiswell R. 1991. The composition of the englacial and subglacial components in bulk meltwaters draining the Gornergletscher, Switzerland. *J Glaciol*. 37(125):59–66.

Valdiya, K.S. 1998. Dynamic of Himalaya. Universities press, Hyderabad. 178.

Vohra, C.P. 1993. Himalayan glacier, Harnessing the eastern Himalayan rivers., Konark Publishers Pvt. Ltd. 120-142.

Wagnon, P., Linda,A., Arnaud,Y., Kumar,R., Sharma,P., Vincent,C., Pottakkal,J., Berthier, E., Ramanathan, A.L. Hasnain, S.I. and Chevallier, P. 2007. Four years of mass balance on Chhota Shigri glacier (Himachal Pradesh, India), a new benchmark glacier in the Western Himalaya, India: *Journal of Glaciology*. 53(183): 603-611.

Chapter 13

Analysis of Rainfall Pattern for Anjar Taluka of Kutch District, Gujarat State

Sanskriti S. Mujumdar and Arvind S. Patel

The rainfall in Kutch District is very scanty and irregular and the district has very few rainy days. The average annual rainfall in the district is between 300 mm to 400 mm. The number of rainy days varies between 10 days to 17 days. Anjar Taluka is located in the central mainland of Kutch district. The region mostly consists of sandy clay soil. It has a geographical area of 121580 hectares. The average annual rainfall for Anjar is 345 mm as against 356 mm average annual rainfall of Kutch district. Through this Chapter it is attempted to study the Rainfall Pattern for Anjar Taluka and compare it with that of Kutch district. For this purpose, annual rainfall data for a period of 130 years from 1878 to 2007 is studied and the probabilities of occurrence and dry and wet spells for the Taluka as well as the region have been determined.

About the Study Area

Kutch is an ancient land possessed of great antiquity, which takes its name from its geographical characteristics and topographical features resembling a tortoise. This crescent shaped region forms part of the northwest Gujarat.

The district stretches roughly from 22°44¢11^{2} (approx.) to 24°41¢25^{2} (approx.) north latitudes and 68°09¢46^{2} (approx.) and

71°54¢47² (approx.) east longitudes. It is bounded on the north and north-west by Sindh (Pakistan), on the north-east by Rajasthan, on the east by the districts of Banaskantha and Mehsana, on the south-east by Surendranagar district, on the south by the Gulf of Kutch and the Rajkot district and on the south-west and west by the Arabian Sea. The district is roughly divided into four parts, *i.e.* Kutch mainland, Wagad region, Island belt comprising of four islands in the desert region and the Rann portion. Rapar Taluka is a part of the Wagad region of Kutch.

The Anjar Taluka is located 23°082N 70°012E/23.13°N 70.02°E/23.13; 70.02. The Taluka of Anjar is situated towards the south east of the mainland of Kutch. It has an average elevation of 81 metres (265 feet). It is situated nearly 10 miles from the Gulf of Kutch. The region is dry and sandy, and entirely depends on well irrigation for its water supply. It includes a town of the same name (Anjar) and 59 main villages in the block. According to the 2001-Census, the total population of the district was 1,526,321 inhabitants. The total population of Anjar city is estimated to be 344,783 or 39.54 per cent of the total population of the district. Anjar experienced additional significant quakes on 21 July 1956 and 26 January 2001.

Rainfall

The normal rainfall for Kutch district ranges between 300 to 400 mm. The district receives an average annual rainfall of 356 mm (for study period of 1878 to 2007), which is erratic and depends on the strength of the summer monsoon. It ranges from 335 mm at Bhuj to 452 mm at Naliya. It varies from 408 mm at Mandvi on the southern coast to 250 mm at Lakhpat on the northwest coast. The variation in the annual rainfall from the year to year is very large. The monsoon rains usually set in by the third week of June in the coastal belt and withdraws by the end of September. The maximum rainfall takes place during the month of July and very less rain occurs during the month of August and September. Most of annual rainfall in the district is received during the southwest monsoon season, July being the rainiest month. Rainfall of about 178 to 468 mm in a single day has also been recorded at many stations.

The average annual rainfall for Anjar taluka is 345 mm. During period 1878 to 2007 the highest annual rainfall amounting to 996 mm for Kutch district occurred in the year 1959 and 1060 mm for

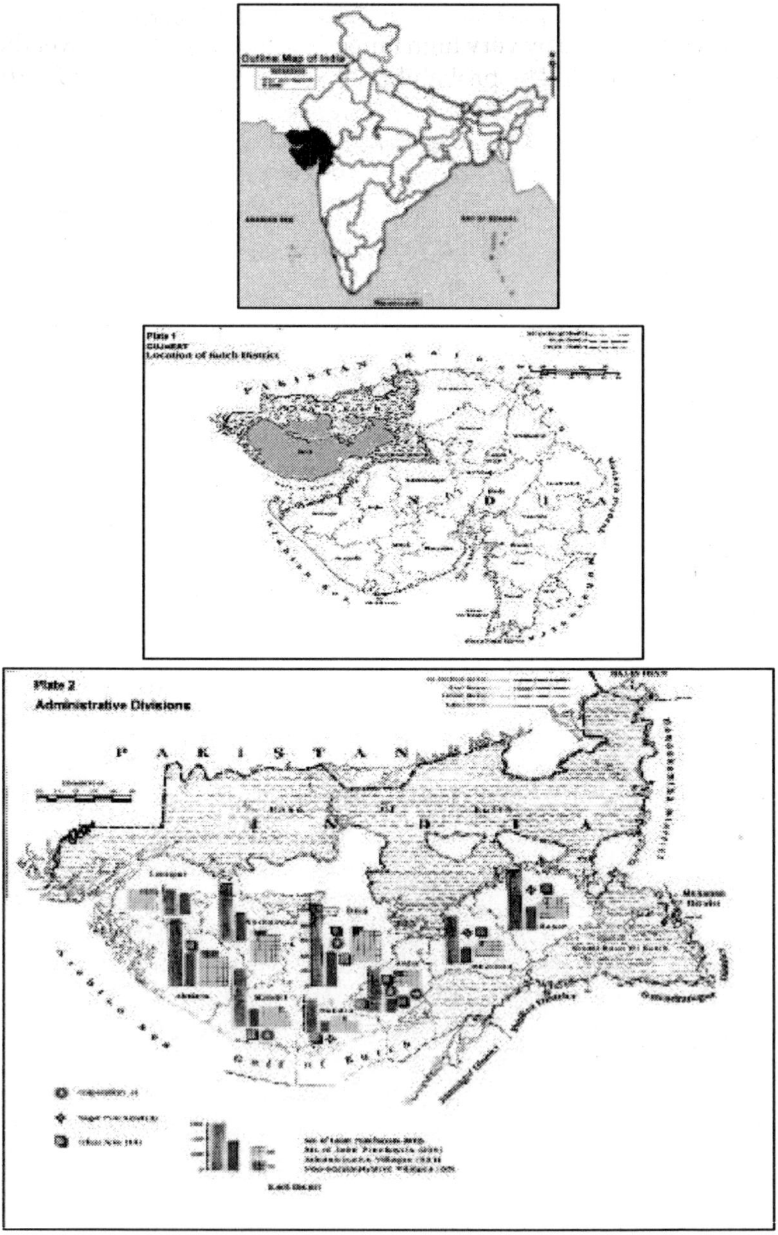

Figure 13.1: Location Map for Kutch District and Bhuj Taluka

Anjar Taluka in the year 1967. Figure 13.2 shows the bar graphs and 3 year and 5 year moving mean graphs for Anjar Taluka as well as Kutch district.

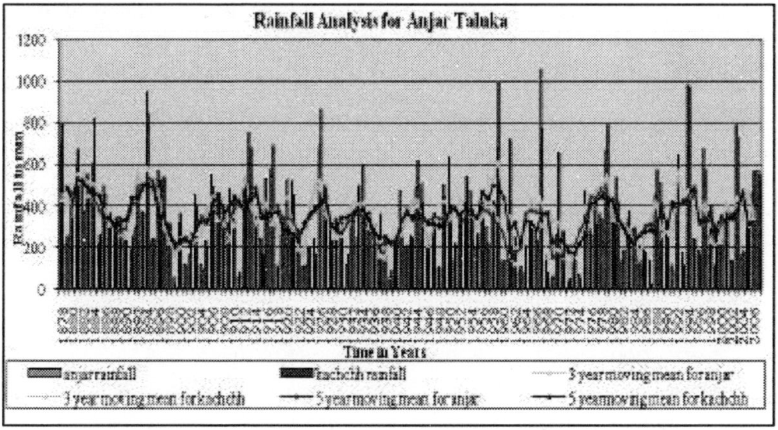

Figure 13.2: Comparision of Average Annual Rainfall of Anjar Taluka and Kutch District

Probability Analysis

Based on the above data, the probability of occurrence of different magnitudes of rainfall has been found out for Anjar Taluka as well as Kutch district. The probability for rainfall occurring between 251

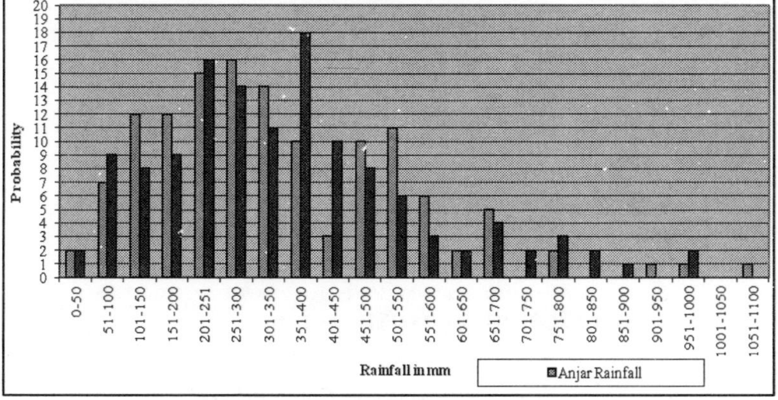

Figure 13.3: Probability of Occurrence of Rainfall for Anjar Taluka and Kutch District

to 300 mm for Bhuj Taluka is maximum while probability of the rainfall deviating from the average annual value by very high range *i.e.* below 50 mm or above 800 mm is very less. The probability of rainfall occurring between 350 to 400 mm for Kutch district is maximum while the probability of rainfall deviating by more than 70 percent is very less.

Table 13.1: Probability of Occurrence of Rainfall For Kutch District

Rainfall in mm	*Number of Events*	*Probability of Occurrence*	*% Probability*
0-50	2	0.0154	1.54
51-100	9	0.0692	6.92
101-150	8	0.0615	6.15
151-200	9	0.0692	6.92
201-251	16	0.1231	12.31
251-300	14	0.1077	10.77
301-350	11	0.0846	8.46
351-400	18	0.1385	13.85
401-450	10	0.0769	7.69
451-500	8	0.0615	6.15
501-550	6	0.0462	4.62
551-600	3	0.0231	2.31
601-650	2	0.0154	1.54
651-700	4	0.0308	3.08
701-750	2	0.0154	1.54
751-800	3	0.0231	2.31
801-850	2	0.0154	1.54
851-900	1	0.0077	0.77
901-950	0	0.0000	0.00
951-1000	2	0.0154	1.54

Surplus and Deficit Year Analysis

The annual rainfall during the study period has been classified as drought, deficit, normal, above average and surplus according to the following criteria as shown in Table 13.3. The analysis has been done for Kutch district in Table 13.4 and Anjar Taluka in Table 13.5.

Table 13.2: Probability of Occurrence of Rainfall for Anjar Taluka

Rainfall in mm	*Number of Events*	*Probability of Occurrence*	*% Probability*
0-50	2	0.0154	1.54
51-100	7	0.0538	5.38
101-150	12	0.0923	9.23
151-200	12	0.0923	9.23
201-251	15	0.1154	11.54
251-300	16	0.1231	12.31
301-350	14	0.1077	10.77
351-400	10	0.0769	7.69
401-450	3	0.0231	2.31
451-500	10	0.0769	7.69
501-550	11	0.0846	8.46
551-600	6	0.0462	4.62
601-650	2	0.0154	1.54
651-700	5	0.0385	3.85
701-750	0	0.0000	0.00
751-800	2	0.0154	1.54
801-850	0	0.0000	0.00
851-900	0	0.0000	0.00
901-950	1	0.0077	0.77
951-1000	1	0.0077	0.77
1001-1050	0	0.0000	0.00
1051-1100	1	0.0077	0.77

Table 13.3: Criteria for Analysis of Rainfall

Sl.No.	*Description*	*Rainfall in mm*
1	Drought years	<150
2	Rainfall deficit years	151-300
3	Normal Rainfall years	300-400
4	Above average rainfall years	400-500
5	Surplus rainfall years	> 500

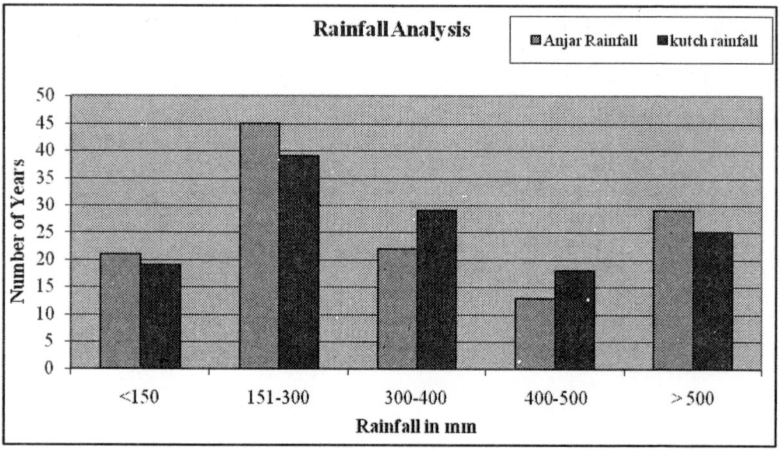

Figure 13.4: Chart Showing the Rainfall Analysis for Anjar Taluka as well as Kutch District

Table 13.4: Rainfall Analysis for Kutch District

Drought Years	*Rainfall Deficit Years*	*Normal Rainfall Years*	*Above Average Rainfall Years*	*Surplus Rainfall Years*
<150	*151-300*	*300-400*	*400-500*	*> 500*
1899, 1901, 1904, 1911, 1918, 1923, 1939, 1948, 1960, 1963, 1968, 1969, 1972, 1973, 1974, 1987, 1991, 1993, 2002	1879, 1885, 1890, 1891, 1895, 1902, 1905, 1915, 1920, 1922, 1924, 1925, 1928, 1931, 1938, 1940, 1942, 1943, 1946, 1951, 1955, 1957, 1962, 1964, 1965, 1966, 1971, 1976, 1982, 1984, 1985, 1986, 1990, 1995, 1996, 1999, 2000, 2004, 2005	1880, 1882, 1887, 1888, 1889, 1893, 1898, 1900, 1903, 1914, 1916, 1919, 1929, 1930, 1934, 1935, 1936, 1937, 1941, 1947, 1950, 1952, 1954, 1975, 1977, 1978, 1980, 1983, 2001	1883, 1886, 1906, 1907, 1908, 1909, 1910, 1921, 1927, 1932, 1933, 1945, 1949, 1956, 1958, 1981, 1989, 1998	1878, 1881, 1884, 1892, 1894, 1896, 1897, 1912, 1913, 1917, 1926, 1944, 1953, 1959, 1961, 1967, 1970, 1979, 1988, 1992, 1994, 1997, 2003, 2006, 2007
Total 19 years	Total 39 years	Total 29 years	Total 18 years	Total 25 years

Table 13.5: Rainfall Analysis for Anjar Taluka

Drought Years *<150*	*Rainfall Deficit Years* *151-300*	*Normal Rainfall Years* *300-400*	*Above Average Rainfall Years* *400-500*	*Surplus Rainfall Years* *> 500*
1899, 1901, 1904, 1911, 1918, 1923, 1931, 1939, 1948, 1951, 1960, 1963, 1964, 1969, 1972, 1974, 1982, 1987, 1991, 1999, 2002	1879, 1885, 1887, 1888, 1890, 1891, 1892, 1895, 1902, 1905, 1908, 1915, 1919, 1922, 1924, 1925, 1928, 1930, 1933, 1936, 1937, 1938, 1940, 1942, 1946, 1955, 1943, 1961, 1962, 1966, 1968, 1970, 1971, 1973, 1984, 1985, 1986, 1990, 1993, 1995, 1996, 1998, 2000, 2001, 2004	1880, 1889, 1893, 1898, 1900, 1907, 1909, 1910, 1914, 1929, 1932, 1935, 1947, 1952, 1957, 1958, 1965, 1976, 1977, 1978, 1983, 2005	1886, 1903, 1927, 1941, 1944, 1953, 1954, 1956, 1975, 1980, 1988, 1992, 2006	1878, 1881, 1882, 1883, 1884, 1894, 1896, 1897, 1906, 1912, 1913, 1916, 1917, 1920, 1921, 1926, 1934, 1949, 1950, 1945, 1959, 1967, 1979, 1981, 1989, 1994, 1997, 2003, 2007
Total 21 years	Total 45 years	Total 22 years	Total 13 years	Total 29 years

Conclusions

For the study period of 1878 to 2007, the Anjar Taluka experienced drought for a period of 21 years as compared to 19 years for Kutch district. There were 45 rainfall deficit years for Anjar Taluka as compared to 39 years for Kutch district. There was normal rainfall for 22 years for Anjar taluka as compared to 29 years for Kutch district. There was rainfall above average annual value for a total of 13 years in Anjar Taluka as compared to 18 years in Kutch district. There was surplus rainfall for 29 years in Anjar Taluka as compared to 25 years for Kutch district. It can be concluded that Anjar Taluka has proportionately more number of drought, rainfall deficit and normal rainfall and excess rainfall years as compared to Kutch district.

References

Mujumdar, S.S. and Patel, A.S. 2010. Study of Ground Water Recharge and its effect on the water quality for Rapar Taluka of Kutch at International Conference on EWRI International Conference held at Chennai between Jan 5-7, 2010.

Mujumdar, S.S. 2001. Effect of Earthquakes on Medium Irrigation Schemes of Kutch District. *M.E. Dissertation* (November 2001).

Mujumdar, S.S. and Patel, A.S. 2008. Impact of Surface Water Storage on quality of Ground Water for Kutch Region of Gujarat at National Conference on *Water Conservation and Artificial Recharge of Ground Water.*

Mujumdar, S.S. and Patel, A.S. 2008. Study of Ground Water Table Fluctuations for Kutch District of Gujarat at National Conference on *Water Conservation and Artificial Recharge of Ground Water.*

Mujumdar, S.S. and Patel, A.S. 2008. Water Resources Management for Arid Region – Kutch A Case Study at National Conference on *Water Conservation and Artificial Recharge of Ground Water.*

Mujumdar, S.S. and Patel, A.S. 2009. Study of Rainfall Pattern and Irrigation Water Utilization for various crops of Rapar Taluka of Kutch at International Conference on *Food Security and Environmental Sustainability* held at I.I.T. Kharagpur between 17-19 December, 2009.

Mujumdar, S.S. and Patel, A.S. 2010. Analysis Of Rainfall Pattern For Bhuj Taluka Of Kutch District – Gujarat State" At International Conference on *Hydrology and Watershed Management (ICHWAM-2010)* held at Hyderabad between 3 - 6 February, 2010.

Mujumdar, S.S. and Patel, A.S. 2010. Study of Ground Water Table Fluctuations and Rainfall Pattern for Rapar Taluka of Kutch at International Conference on EWRI International Conference held at Chennai between Jan 5-7, 2010.

Report on "Strategic Research and Extension Plan of Kutch District." 2006. Submitted by Agricultural Technology Management Agency of Kutch District.

Reports of District Statistical Officer, Kutch.

Reports of Gujarat Ecology Commission.

Reports of Kutch Irrigation Circle.

Surendra, Singh and Amal Kar (1996) "Integrated Natural and Human Resources Appraisal for Sustainable Development of Kutch District", CAZRI.

Sutaria, P.N., Patel, A.S. and Mujumdar, S.S. 2002. Drinking water management for summer season in an earthquake affected severely drought prone region- Kutch: A Case Study – a paper presented in One Day State Level Seminar on Sustainable Water Management organized by Indian Water Works Association, Vadodara Chapter.

Chapter 14

Impacts of Climate Change on Water Resources and Agriculture in Rajasthan

Nidhi Kaushik

Introduction

Climate Change is one of the major problems confronting the world today. Data on temperature since 1000 AD shows that the 20th century was unusually warm and the decade of 1990's was the hottest with six warmest years occurring within this decade. Climate change is manifesting itself in many ways across the country. The Intergovernmental Panel on Climate Change (IPCC) in its fourth assessment report observed that, "warming of climate system is now unequivocal, as is now evident from observations of increases in global average air and ocean temperatures, widespread melting of snow and ice and rising global sea level" (Solomon *et al.*, 2007). Changes in surface temperature, rainfall, evaporation and extreme events since the beginning of the 20th century have been observed. The atmospheric concentration of carbon dioxide has increased from about 280 parts per million by volume (ppmv) to about 382 ppmv and the global temperature of the earth has increased by about 0.6°C. The global mean sea level has risen by 10 to 20 cm. There has been a 40 per cent decline in Arctic Sea ice thickness in late summer to early autumn in the past 45–50 years.

India is a large country with 15 agro-climatic zones, with diverse seasons, crops and farming systems. For a majority of people in

India, to this day, agriculture is the main source of livelihood. Agriculture is the most vulnerable sector to CC as it is inherently sensitive to climate variability and CC will leave its impacts on Indian agriculture in various direct and indirect ways. This obviously means an impact on the lives and livelihoods of millions of Indians.

For instance, it is reported that about two-thirds of the sown area in the country is drought-prone and around 40 million hectares is flood-prone. The poorest people are likely to be hardest hit by the impacts of climate variability and change because they rely heavily on climate-sensitive sectors such as rainfed agriculture and fisheries. They also tend to be located geographically in more exposed or marginal areas, such as flood plains or nutrient-poor soils. The poor also are less able to respond due to limited human, institutional and financial capacity and have very limited ability to cope with climate impacts and to adapt to a changing hazard burden.

Rajasthan, located in the northwest part of the country, is the largest state in India with a geographical area of 3, 42,239 sq. km. Rajasthan has 5.5 per cent of the country's population (56.51 million as per Census 2001), of which 77 per cent reside in rural areas. The state has four major physiographic regions- the Western desert, the Aravalli hills, the Eastern plain and the South- Eastern plateau. The climate of the state varies from semi-arid to arid. The key sectors in which Rajasthan already faces challenges (due to pressures from a growing population, rapid economic growth and degradation of environment and natural resources) that are likely to exacerbate under conditions of change in the climate.

On top of the above challenges, the State also falls within the areas of greatest climate sensitivity. In an era of climate change, Rajasthan is likely to suffer further water shortage due to overall reduction in rainfall. In addition, the State has the maximum vulnerability and lowest adaptive capacity to climate change challenges. Rajasthan has the maximum probability of occurrence of drought in India. Condition may deteriorate in terms of severity of droughts in Rajasthan. Even though 20 per cent rise in all-India summer monsoon rainfall is projected, in Rajasthan overall rainfall is projected to decrease, and evapotranspiration to increase, due to global warming.

Even 1 per cent increase in temperature from base data could result in an increase in evapotranspiration by 15 millimeter (mm), resulting into additional water requirement of 34.275 million cubic meter (mcm) for Jodhpur district alone and 313.12 mcm for entire arid zone of Rajasthan. Although analysis of 100-year rainfall data for the arid region of Rajasthan indicates an increasing trend of 0.5 mm/year, but increased evapotranspiration demand due to global warming can put tremendous pressure on existing overstressed water resources of this region. As the total available surface water resources of the arid zone are to the tune of 1361.21 mcm, a robust policy intervention is required for sustainable water management. Studies have documented a rising trend in temperature at Barmer, Jodhpur, Ajmer and Pali in Luni river basin of arid western Rajasthan. In the same region, annual rainfall has shown increasing tendency at 19 stations (around Ajmer in upper part of the Luni basin). Decreasing temperature trends have been observed at Udaipur and Jwaibandh, and decreasing rainfall trend at the remaining nine stations in lower Luni basin, *i.e.*, Barmer. Overall, there is a prediction of an increase in the rainy day intensity by 1–4 mm/day in India, but some areas in the northwest India the rainfall intensities are predicted to decrease by 1 mm/day.

Widespread land degradation is a persistent challenge in Rajasthan. Recent studies have further predicted that due to climate change there may be significant increase in the desert area over India in next 100 years with potentially disproportionate impact of global warming on coupled human and natural systems. Model studies on the wind erosion potentials in the Thar region for AD 1951 to 2100 suggest that larger efforts in land-conservation practices would be required than at present to stabilize the aeolian bed forms in the Thar Desert.

Impacts on Water Resources

Climatic changes are likely to impact significantly upon water resources availability. In India, demand for water has already increased manifold over the years due to urbanization, agriculture expansion, increasing population, rapid industrialization and economic development. Rajasthan covers nearly 10.5 percent of the country's geographical area, with surface water resources in the state being only 1.16 percent of the country's available surface water. It is the driest state in India with nearly two-thirds of the land area classified as arid and semi arid. The average annual rainfall in the state is about 531 mm (about 318 mm in the desert areas) as compared to the national average of 1200 mm. There is only one perennial river in the state *i.e.* Chambal river. Water resources in the state are not only scarce but also have a highly uneven distribution both temporally and spatially with most of it confined to the south and southeastern parts of Rajasthan. The total surface water in the state is 21.71 BCM, out of which 16.05 BCM is economically utilizable. The state has so far harnessed 72 per cent of the economically utilizable portion. There exists a huge deficit between water demand and supply in Rajasthan, nearly 8 BCM currently, which is likely to increase to 9 BCM by the year 2015. Droughts are rampant in Rajasthan. She Wale et al (2001) estimated the percentage area in India affected by moderate and severe drought and found (based on data analysis for a 124 year time-period, 1875- 1998) that the probability of occurrence of droughts was maximum in West Rajasthan. The probability of moderate drought in Rajasthan was found to vary between 17- 24 per cent, and between 2-14 per cent in case of severe drought.

Rajasthan is principally dependent on groundwater for meeting nearly 90 percent of its drinking water needs. Scanty and irregular rainfall coupled with over-exploitation of the groundwater sources has resulted in poor groundwater recharge as compared to its extraction for various purposes, leading to a decline in groundwater levels across the state. The ratio of ground water extraction to its annual replenishment was 35 per cent in 1984 and 138 per cent in 2008. Geological factors combined with over -extraction have brought adverse changes in the geochemistry of groundwater in Rajasthan, often concentrating naturally occurring contaminants. Natural contaminants such as fluoride, nitrate and chloride salts have been found to be increasing in ground water in different parts of Rajasthan, making it unfit for drinking and posing serious health risks to communities extracting water from these contaminated aquifers.

The water resources in Rajasthan are not adequate to meet the rising water demands from the domestic, agriculture and industrial sectors; and climate change will act as an additional stressor to water availability in the state. Studies conducted as part of India's first National Communications (NATCOM) to the United Nations Framework on Climate Change (UNFCCC) indicate that some river basins in India are likely to face an overall reduction in the quantity of the available runoff under a climate change scenario (increased

GHG), projected for the period 2041- 2060. The west-flowing rivers of Kutch, Saurastra and Luni which occupy about 60 per cent of the area in Rajasthan are likely to face conditions of acute water stress. The river basin of Sabarmati is likely to experience constant water shortage. Increase in evapo transpiration for the state of Rajasthan has been identified as one of the key impacts of climate change on water resources.

Impact on Agriculture and Animal Husbandry

The agriculture sector contributes 27 per cent to Rajasthan's Gross Domestic Product (GDP) and is critical to the economy of the state. The major agricultural land use is rainfed cropping with pearl millet being the main cereal crop along with kharif legumes (clusterbean, moth bean, mung bean). Eastern Rajasthan falls in the semi-arid, 500–1000mm annual rainfall zone and is intensively cultivated for pearl-millet/sorghum/kharif pulses/maize wheat/ barley/mustard/rabi pulses. Agriculture in Rajasthan is likely to face several challenges in the light of increasing competition for resources such as water, land and energy, from nonagricultural sectors, along with increasing food demand due to the rising population, increased use of fertilizers due to increased production demand and increased water withdrawals for irrigation from canals, wells and tube wells.

The contribution of animal husbandry to the state GDP is about 9.16 per cent. Rajasthan ranks second highest in milk production in the country (amounting to nearly 17 lakh kg per day). More than 80 per cent of the rural families in Rajasthan keep livestock in their households. This sector is a source of self employment to majority of the households in rural and urban Rajasthan and supplements their income and nutritional requirements. About 35 per cent of the income of small and marginal farmers in the state is derived from animal husbandry, with this share being as high as 50 per cent in the arid regions. Lack of proper feed and fodder and poor market linkages however, affect productivity in this sector. This sector has a high potential to create employment in rural areas with lesser investments as compared to other sectors.

Agriculture will be adversely affected not only by changes in the overall amounts of rainfall, but also by shifts in the timing of the rainfall. Changes in climatic variables such as temperature and precipitation, increased incidence of droughts and intensification of water- scarce conditions can adversely impact state agricultural production and productivity owing to changes in the hydrological cycle, changes in plant physiology, soil properties (including nutrient availability), rate of evapo-transpiration, pest dynamics and rate of weed proliferation. Changes in precipitation patterns can also increase the likelihood of short-run crop failures and production declines. Climate change poses formidable challenges to the animal husbandry sector as well. Though the vulnerability of farm animals to climate change varies with their genetic composition, type and breed, life stage and nutritional status, studies unambiguously indicate that the performance of farm animals is sensitive to the climate. Heat stress in dairy animals is likely to impact their productive and reproductive performance. Reduction in feed and fodder availability due to changes in the climate can also affect livestock productivity. During droughts, livestock play a larger role in supporting the income and sustaining the rural population. However, under conditions of declining water and fodder availability, the contribution of this sector to household incomes of farming communities declines drastically.

Conclusion

The current simulation results from GCMs(Global Climate Models) are still considered uncertain. Ability of present GCMs in predicting the impact of climate change on rainfall is still not promising. In addition, there are uncertainties involved in predicting extreme flood and drought events by these models. This Chapter summarizes some important vulnerability issues associated with the present and potential future hydrological and agricultural changes due to climate change. The National Environment Policy also advocated that anthropogenic activities have severe adverse impacts on India's precipitation patterns, ecosystems, agricultural production, forests and water resources. Large-scale planning would be clearly required for adaptation measures for climate change impacts, if catastrophic human misery is to be avoided. Every year some parts of Rajasthan are affected by drought. But the State has failed so far to diagnose the drought phenomenon and to come up with a long term solution. Presently, Climate-related changes have brought widespread misery and huge economic losses to India, adversely affecting public health, food security, agriculture, water resources and biodiversity. The situation is likely to worsen if human beings continue to pump 'greenhouse gases' (GHGs) like carbon dioxide into the atmosphere. Specifically, population, income, and economic growth could all affect the severity of climate change impacts in terms of food security, hunger, and nutritional adequacy. If climate change adversely affects agriculture, human effects are likely to be more severe in a poorer world with more people near hunger. In a prosperous economic environment, consumers may suffer a greater economic loss but be less likely to suffer the chronic effects of hunger.

References

Brenkert, A.L. and E.L., Malone. 2004. Modeling Vulnerability and Resilience to Climate Change: A Case Study of India and Indian States. *Climatic Change*. 72(1):57-102.

Kumar K.V. 2007. Climate Change studies in Agriculture, Economic and Political Weekly, November 17, 2007.

Kumar, K.S. Kavi. 2003. Vulnerability of Agriculture and Coastal Resources in India to Climate Change. Research report submitted to Ministry of Environment and Forests, GoI, New Delhi.

Mall, R.K., Gupta, A. Singh, R., Singh, R.S. and Rathore, L.S. 2006. Water resources and climate change: An Indian perspective. *Current Science*. 90:1610-1626.

Goyal, R.K. 2004. Sensitivity of evapotranspiration to global warming: a case study of arid zone of Rajasthan (India), *Agricultural Water Management*. 69:1-11.

Singh, R.S., Narain, P. and Sharma, K.D. 2001. Climate changes in Luni river basin of arid western Rajasthan (India). *Vayu Mandal*. 31:103–106.

Kumar, K.S. Kavi and Parikh, J. 2001a. Socio-economic Impacts of Climate Change on Indian Agriculture. *International Review of Environmental Strategies*. 2(2):277-93.

Pandey, D.N., Gupta, A.K. and Anderson, D.M. 2003. Rainwater harvesting as an adaptation to climate change. *Current Science*. 85:46-59.

Sharma, K.D. 2009. Groundwater management for food security, *Current Science*. 96: 1444-1447.

Rodell, M., Velicogna, I. and Famiglietti, J.S. 2009. Satellite-based estimates of groundwater depletion in India. *Nature*. 460: 999-1002.

Narain, P., Khan, M.A. and Singh, G. 2005. Potential for Water Conservation and Harvesting against Drought in Rajasthan, India, Working Paper 104/Drought Series: Paper 7, International Water Management Institute, Colombo, Sri Lanka.

Panwar, N.L. and Rathore, N.S. 2009. Potential of surplus biomass gasifier based power generation: A case study of an Indian state Rajasthan, Mitigation and Adaptation Strategies for Global Change. 14:711-720.

World Wildlife Fund (WWF). 2005. An Overview of Glaciers, Glacier Retreat and Subsequent Impacts in Nepal, India and China. WWF Nepal Programme (March 2005).

Chapter 15

Impact of Thuthi-Kankasiya Check Dam on Irrigation in Machhan River Basin

N.J. Shrimali and A.S. Patel

Water is a key element and a very important natural resource which forms the basis of all life. It is very important tool for human development, and resource in all economic activities ranging from agriculture to industry. With ever increasing pressure of human population, there is severe stress on water resources. The finite nature of renewable fresh water makes it a critical natural resource to examine in the context of population growth. In recent past a significant awareness in the field of conservation of water resources has been observed. Check-dams have been the heroes of watershed development programs everywhere. Check-dams are invariably down-river, down-in-the-valley structures where land is already quite productive and comparatively well served with water. Many NGOs have been actively engrossed in the field of environment and water resources particularly by the social aspects of that field. N.M. Sadguru Water and Development Foundation (Dahod) is one such organization working in this field. They have constructed 10 check dams in series on the river Machhan Nala, a small tributary of the river Anas, located in Jhalod Taluka of Dahod district of Gujarat. These check dams are built to harness the entire river course and recharging. Lift irrigation is also practiced from the storage created by these check dams.Thuthi-Kankasiya is the last check dam constructed in this series just upstream of meeting of Machhan with Anas.

Introduction

Gujarat State is situated between 20° 06' and 24° 42' North Latitudes and 68° 10' and 74° 28' East Longitudes in the western part of India. The State has the longest coastline in the country measuring about 1,600 km along the western part of India. The rainfall in the state is moderate. Climatic conditions are varied in nature. The climate in general has three main seasons *i.e.* summer, monsoon and winter. The overall climate is humid, sub-humid and semi-arid to arid.

In recent past a significant awareness in the field of conservation of water resources has been observed. Check-dams have been the heroes of watershed development programs everywhere. Check-dams are invariably down-river, down-in-the-valley structures where land is already quite productive and comparatively well served with water. It is always the upper reaches in undulating terrains that are harder to farm and therefore left to the marginal people.

Many NGOs have been actively engrossed in the field of environment and water resources particularly by the social aspects of that field. N.M. Sadguru Water and Development Foundation (Dahod) is one such organization working in this field. They have constructed 10 check dams in series on the river Machhan Nala, a small tributary of the river Anas, located in Jhalod Taluka of Dahod district of Gujarat. Thuthi-Kankasiya is the last check dam constructed in this series just upstream of meeting of Machhan with Anas.

Study Area

The study area is located in Jhalod taluka of Dahod district. The nearest railway station to reach the taluka place is Dahod, which is about 35 km from Jhalod. Jhalod is connected with district head quarter Dahod by State Highway-58.Jhalod is also connected by road with Kushalgadh and Banswada the prominent places of Rajasthan.

Dahod district is moderately humid area. The minimum humidity observed in the month of March and April. The sunshine hours ranges between 11 hours to 4 hours. The various seasons of the year are:

1. Monsoon: (Middle of June to middle of October)

2. Winter: (November to February) and
3. Summer: (March to June).

From March onwards the temperature starts rising till it reaches the maximum, as high as 45°C January is the coldest month of the year.

Machhan River is flowing from South-West to North-East direction. About 1300 rivulets meets and helps to get collection of rain water into the river. The overall topography of the region from which Machhan river flows is highly undulating and of varying slopes. Machhan river catchment receives much of its rainfall from the southwest monsoon during the period between June and September, its maximum intensity being in the months of July and August. The annual rainfall in the area varies between 800 to 1000 mm.

Check Dams on Machhan River

Totally 10 check dams have been constructed on river Machhan in series. They are as follows

1. Kachumber Falia Check Dam
2. Wankol Check Dam
3. Simaliya Check Dam
4. Therka Check Dam
5. Kharsana Check Dam
6. Bambela Check Dam
7. Mahudi Check Dam
8. Rajudia Check Dam
9. Kheda Ambazaran Check Dam
10. Thunthi Kankasiya Check Dam

Thunthi Kankasiya check dam is located at downstream most part of river Machhan at village Thunthi Kankasiya. It is on just upstream of the location where Machhan meets Anas near the border of Gujarat and Rajasthan. It was constructed by N.M. Sadguru Water and Development Foundation (NMSWDF) in 1994. It was constructed in 159 days time. Following Table 15.1 gives salient features of it.

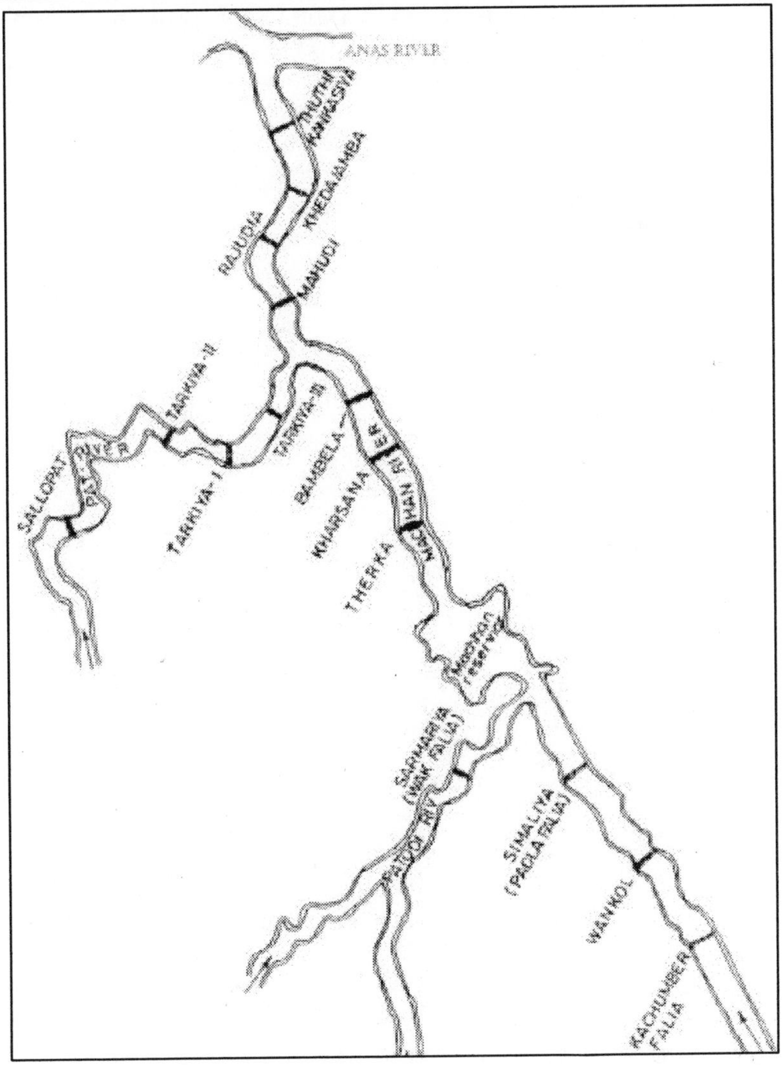

Figure 15.1: Check Dams on River Machhan

Analysis

It is observed that the area irrigated variation follows pattern of rainfall variation. Year 1994 was year of one of the highest rainfall 1350 mm. In this year irrigation from four lift irrigation schemes was

Table 15.1: Salient Features of Thuthi-Kankasiya Check Dam

Year of Construction	*1994*
C.A. (Mile^2)	44.5
Area Irrigated (Acres)	350
Total Length (m)	105.5
Length of Crest (m)	75.5
Bottom Width (m)	5
Height (m)	3.5
Capacity (MCFT)	16
Latitude	23º10'27"
Longitude	74º13'58"

Table 15.2: Rainfall Characteristics–Thuthi-Kankasiya Catchment

Year	*Rainfall (m.m.)*	*Deviation from Last 10 Years (m.m.)*	*Deviation from Last Year (m.m.)*	*Year*	*Rainfall (m.m.)*	*Deviation 10 Years (m.m.)*	*Deviation from Last Year (m.m.)*
1992	549	-169.855	-101	2000	231	-514.83	-121
1993	712	-7.155	163	2001	536	-118.4	305
1994	1124	415.845	412	2002	722	79	186
1995	392	-352.055	-732	2003	1015	354.7	293
1996	869	126.645	477	2004	1125	434.4	110
1997	950	165.55	81	2005	463	-227.7	-662
1998	715	-104.05	-235	2006	1154	456.2	691
1999	352	-444.73	-363	2007	736	9.7	-418

practiced. In 1995 this number increased to 6 in spite of fact that the rainfall was very less, less than 50 per cent of the previous year. Up to 1998 irrigation is found to have been practiced from L.I. Schemes at 6 places though rainfall was varying from as high as 1258 mm in 1996 to 828 in 1997 and 1037 mm in 1998. In 1999 rainfall was less, 630 mm and irrigation could be practiced from 5 LI schemes.

The check dam was constructed in 1994. From 1996 to 1997 the rainfall was continuously decreasing even though there observed

Table 15.3: Area Irrigated by Thuthi-Kankasiya Check Dam

Year	*Rainfall (mm)*	*Area Irrigated (Sq. Km.)*
1994	1375.8	0.49
1995	626	0.3
1996	1270	0.47
1997	814	0.54
1998	1012	0.55
1999	663	0.12
2000	257	0
2001	488	0
2002	481	0.12
2003	716	0.38
2004	829	0.48
2005	672	0.7
2006	1631	0.38

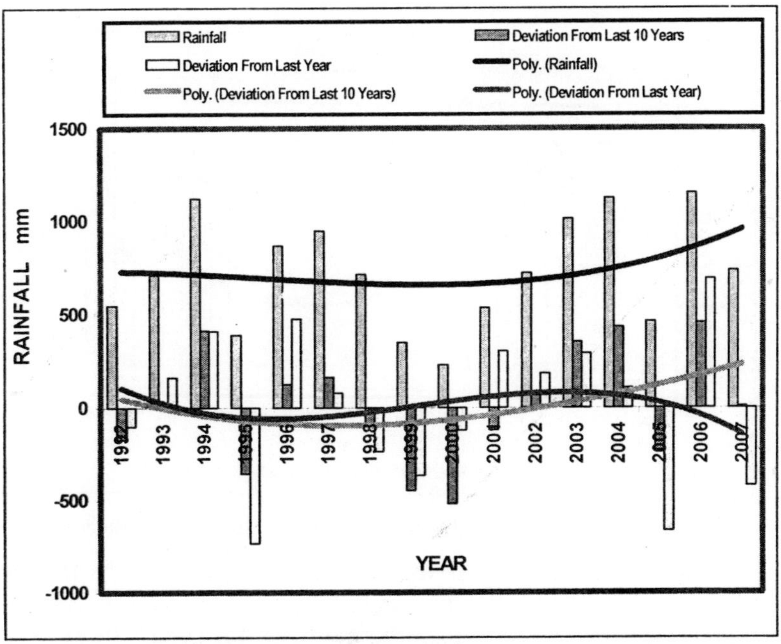

Graph 15.1: Rainfall Characteristics–Thuthi-Kankasiya Catchment

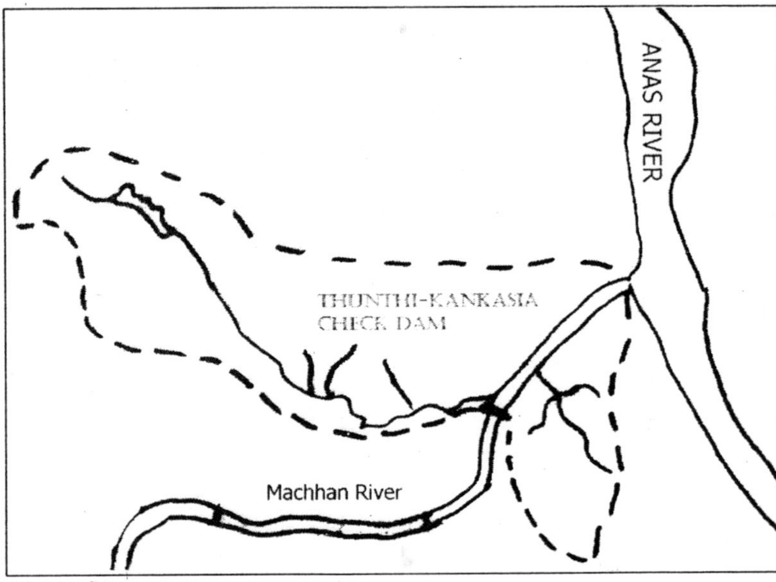

Figure 15.2: Catchment of Thuthi-Kankasiya Check Dam

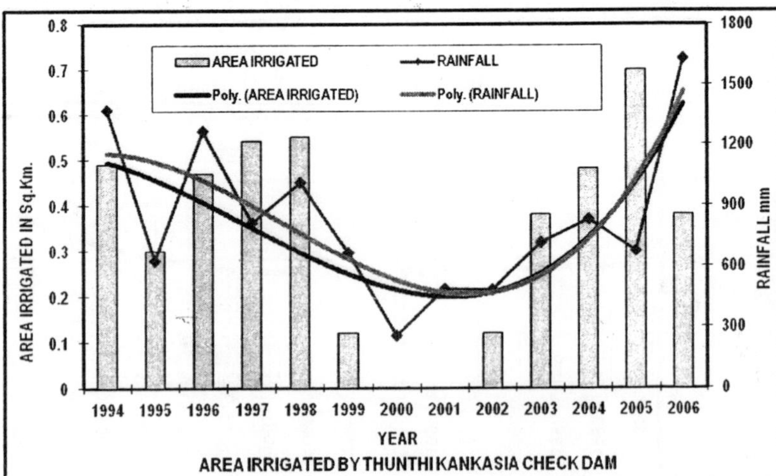

Graph 15.2: Area Irrigated by Thuthi-Kankasiya Check Dam

rise in irrigation done. Years 2000-2002 were the years of severe drought. Because of this no irrigation could be done in 2000 and 2001. However in spite of less rainfall of same magnitude as in 2001 some irrigation was done in 2002.

Discussion on the Analysis and Conclusions

The rainfall in the region of study area is highly erratic. In this region rainfall varies from 580 mm to 800 mm annually. There is high variability in the rainfall with the coefficient of variation from 37 to 65 per cent. It is observed that rainfall is unevenly distributed over the area. Further it can also be observed that the average annual rainfall of the decade is having least variation but the rainfall of any individual year is highly varying. Study of rainfall data indicates that during 8 years out of 13 years annual rainfall in the study area are found less than average rainfall. Rainfall is highly erratic.

Irrigation is practiced from the storage of check dams by Lift Irrigation Schemes. Area irrigated mainly depends upon amount of storage which in turn depends on rainfall. It is worth noticing that sustained irrigation is achieved in less rainfall years. During Year 2002, year of third consecutive severe drought some amount of irrigation has been noticed which shows recharge effect due to storage created by check dam and increased base flow.

References

Agriculture Report, Machhan Nala Irrigation Project.

Merh S. S. *Geology of Gujarat*. Geological Society of India, Bangalore.

Mukherjee, Sanjeeb. 2003. Impact Of Check Dams In Groundwater Recharge-A Case Study Of Machhan River Basin. *M.E. Dissertation*. The M.S. University of Baroda.

Patel, A.S.and Shah, D.L. 2008. *Water Management – Conservation, Harvesting and Artificial Recharge*. New Age International Publishers (1-2008).

Shrimali N.J. 2009. Evaluation of irrigation Cum Artificial Recharge Structures In Machhan River Basin Through Mathematical Model. *Ph.D. Thesis*. M.S. University of Baroda, Vadodara.

Part-IV

RESPONSE TO WATER CRISES

Chapter 16

Water Insecurity: The Plachimada Struggle

Raju Narayana Swamy

The earth, with its varied and copious life forms, including over six billion humans, is facing a serious water crisis. This crisis is one of water governance, essentially caused by the ways in which we mismanage water. But the real heartbreak is the effect it has on the everyday lives of poor people, who are shattered by the burden of water-related disease, living in degraded and often dangerous environments, struggling to earn a living, and to get enough to eat. The crisis is experienced also by the natural environment, which is groaning under the mountain of wastes dumped onto it daily, and from overuse and misuse, with apparently little care for the future generations. The security of individuals, communities, nations, and the entire global community is increasingly jeopardized by unpremeditated, non-military environmental threats. These threats are self-generated. The United Nations has recognized access to water as a basic human right, stating that water is a social and cultural good, not merely an economic commodity. Today, due to increasing consumption patterns water is becoming scarce and this scarcity is an emerging threat to the global population. Global consumption of water is doubling every 20 years, more than twice the rate of human population growth. At present more than one billion people on earth lack access to fresh drinking water. By the year 2025 the demand for freshwater is expected to rise to 56 per cent above what currently available water can deliver if current trends persist. We continue it on ourselves, by fouling our air and water, and over harvesting our land. Despite these Omni present connections, environmental issues are still not high on the national security agenda. The Chapter examines water privatization in the context of national water policy- using the experience gained from the case study-Plachimada. Plachimada village of Perumatty panchayat in Chittoor taluk is a small hamlet in Palakkad district in Kerala. The site of the factory is located barely three kilometres to the north of the Meenakkara Dam reservoir and a few hundred meters west of the Kambalathara and Venkalakkayam water storage reservoirs. The Moolathara main canal of the Moolathara barrage passes less than ten metres north of the factory compound and the main Chittoor River runs very close to the Coca Cola plant. The Plachimada ground water dispute may be seen as a microcosm of what many believe is a coming century of global water conflict. In this paper, we argue that privatization of water is a violation of basic rights of citizens and that privatization creates water monopolies and brings about more disparity in access to water. The study examines the impact of Coca cola factory causing the destruction of ground water and the impact of these activities on the people. The human activities that are responsible for the water crisis have been extensively detailed. The problem is compounded by the growing interest in commercialisation of water resources and services. The corporate interests in water and the global nexus of big business lobby organisations, and for-profit water corporations to which it belongs is been analysed with.

Introduction

Right to safe drinking water is basic for livelihoods. Financial, technological, institutional as well as societal factors hold back access to safe drinking water. The poor may not consume as much water and often the marginal productivity per unit of water/whether measured in terms of good health, longevity or income- is much greater for the poor than for those who are more affluent. This is because the access of the poor to safe water is so limited to begin with that even a slight improvement in the quality or quality is likely to have a far more noticeable impact on their lives and livelihoods. Therefore the term "water Security" refers to the vulnerability of economies and ecosystems and also individuals and communities, to the escalating stresses on the World's limited water resources (World Water Council, 2000; Nitish Jha, 2010).

Traditionally, India has been well gifted with large freshwater reserves. But the over exploitation of surface and groundwater over the past few decades has resulted in water scarcity in many areas. Localised water depletion, resulting in falling of water levels in the immediate vicinity of a well or other sources in rural India will be a crisis. This is most likely to occur where the demands being placed on a ground water sources are high. Increased stress on a groundwater source during drought seasons also makes the problem more severe. In India water stress is a much more localised problem than food insecurity. Required amounts of food can be transported further and more easily than those of water in areas where there are no piped networks. The International Water Management Institute forecasts that by 2025, thirty three per cent of India's population will live under absolute water scarcity condition. The per capita water availability, in terms of average utilizable water resources in the country, has dropped drastically from 6008 M3 in 1947 to 2200 M3 by 2025. The requirement of water by different sectors by 2025 is estimated to be 105 Mha.m but the share of water for agriculture is expected to reduce fro the present level of eighty five per cent to seventy one per cent by 2025 (Kumara Charyula *et al.*, 2007).

India's Water Policy Framework

Water is a natural resource that originally was made freely available by nature. The role of Welfare state was to respect such traditional management systems and protect the natural resources from over exploitation by other actors. However the new liberal economic order has introduced changes in the role of the Indian state and forced it to relinquish its protectionist role in the economy and development of the country and make way for the market forces to undertake their economic activity (Sanjay Gathia, 2005). The first National Water Policy was adopted by the National Water Resources Council in 1987. Since then a number of issues and challenges have emerged in the progress and administration of water resources. In the last few years a number of problems/challenges have emerged in the development and management of water resources. On the one hand the after crisis (of its reduced availability per head and deteriorating quality) has grown and on the other hand local initiatives in water conservation have started to respond to the challenge strengthening there (Kathpalia, G.N. and Rakesh Kapoor, 2004). On 1st April, 2002 the National Water Resources Council met under the chairmanship of the Prime Minister and adopted the National Water Policy 2002, a revised version of the earlier policy of 1987.

The new policy recognized water as a precious national asset. It embodies the nation's resolve that planning, development and management of water resources could be governed by National Perspectives. The new policy integrated water resources development and management for optimal and sustainable utilization of the available surface and ground water, creation of well-developed information system, use of traditional methods of water conservation, non-conventional method for water utilization and demand management. Greater emphasis has been laid on water quality aspects. The policy belies the people's expectations of decentralization of water development and management and encouragement of community-based water harnessing. The policy has left many loopholes for the market forces to take advantage. It does not have any punitive mechanism against wrong doers on a clause for encouraging those who harnessed water (Bansil, 2004). The new policy holds that all sources of water, rainfall, river water, surface ponds, lakes and ground water are indivisible and part of a larger ecological system. It also recognises that water is necessary for sustaining all forms of life. This recognizes that water is essential for realizing the right to life and livelihood. The policy again recognizes that the existing water resources would have to meet the needs of drinking water, household purposes and also the needs of the livestock.

The national water Policy 2002 has also stressed on the private sector participation in the policy. Private sector participation should be encouraged in planning development and management of water resources projects for diverse uses, wherever feasible. Private sector participation may help in introducing innovative corporate management and improving service efficiency and accountability to users. Depending upon the specific situation, in building, operating, leasing and transferring of water resources facilities, may be considered.

Source: Bansil, P.C. (2004). Water Management in India, Concept Publishing Company, New Delhi, p.518.

Coca-Cola in Kerala

Plachimada village in the Palakkad district is located in Kerala, the southern most part of India. The Hindustan Coca-Cola Beverages Pvt. Ltd established a unit in 1998-99 in a 40 acre plot at Plachimada. The factory is located a few metres away from the main irrigation canal form the Moolathara barrage. The area is a predominantly agricultural area. Paddy being the dominant crop, sugar cane is also planted in some places. The main source of water for the villagers is the canals from the Malampuzha Reservoir which irrigates thousands of acres of fields. But water has been in short supply in these villages. The adivasis living here are mainly agricultural labourers. They had no land and no means to survive as the traditional ways of life came to a standstill because of massive drop in paddy cultivation. It was then the global giant came with their Cola factory to their village. Ecosystem people depend to a very large extent on natural resources in their immediate surroundings, the local ecosystem, where as omnivores have the power to draw on resources from a wide, sometimes even global, catchments area (Eva, Wramner, 2004). The Hindustan Coca-Cola Beverages was given the license to establish its unit in 1999 for producing the Coca-Cola brand of beverages including bottled mineral water. 200 truck loads of beverages containing 550-600 cases were produced everyday. For this 600,000 litres of ground water are extracted from 65 bore wells dug on the land. The water level in the wells and other sources dropped in an unusual manner and the village turned into a kind of a desert within a few months.

Water Power and Resistance

The first major disagreement that erupted between the local people and the Company was regarding the pumping of water which was causing scarcity of drinking water in the village, and its pollution by factory effluents. Plachimada lies in a rainshadow area and the amount of withdrawal here is always in excess of the recharge component from the rainfall, and hence it is the excessive groundwater pumping from a "rainshadow" and "overexploitation" region that is the real issue in Plachimada (Ravi Raman, K., 2010). On April 7th 2003, the village panchayat of the area took a decision to cancel the license agreement granted to the company. Under the Kerala Panchayat Raj Act, the local panchayat has authority to issue or suspend a license or impose conditions on an industry. This decision was challenged by the Company in the High Court. On 16th December 2003, the Kerala High Court directed the Company to find out alternative sources of water and ordered the Panchayat not to interfere with the working of the Company if it used such alternative sources. It also directed the Company to use only that quantity of water which was available to the owner of 34 acres of land.

> In India 69 per cent of people in non-irrigated areas are poor; in irrigated areas this figure falls to 26 per cent. See also *Water for People Water for Life*, World Water Assessment Programme, United Nations World Water Development Report, p.18.

Over extraction of ground water by Coca Cola company violates three International legal regimes, American Common Law Reasonable Use Rule, the Restatement (second) of Torts Reasonable Use and Doctrine of Correlative Rights, governing the use of groundwater. Also Article 21 of Indian Constitution, in which right to water is a fundamental right.

On January 15, 2005 the local community people demanded the permanent closure of the factory. The Coca Cola Virudha Samithy (Anti Coca Cola people's Struggle Committee) and the Plachimada Solidarity Committee are the two strong forces that lead the movement against Coca Cola in Plachimada. The residents of Plachimada have been on the vigil by 24 hours a day outside the gates of the Coca Cola factory in their village. The vigil has crossed the mark of 1280 days on October 22, 2005. People representing the five most affected colonies adjacent to the factory, who belong to Eravala, Malasar tribal communities and other scheduled caste communities have been holding demonstrations and sit-in strike in front of the factory. They picketed the Hindustan Coca-Cola Beverages Pvt. Ltd. At Plachimada and gave an ultimatum to the authorities to quit immediately. In March 2005, the Perumatty Panchayat refused to renew the license issued to Cola Company forcing it to shut down its operations (Ashish Kothari and Anuprita Patel, 2006). The Cola Company again appealed to High Court on April 7, 2005 gave a further ruling allowing the company to extract 500,000 litres of water per day, under normal rainfall conditions. The Kerala State Pollution Control Board, on August 19, 2005, directed the Coca Cola plant in

Plachimada to stop making of all kinds of produce with urgent effect because of high levels of cadmium around the plant.

These water movements has already had a profound effect on global water politics, forcing global institutions such as the World Bank and the United Nations to address the inadequacies of their policies, and has helped formulate water policy inside countries. The growth of a democratic water justice movement is a critical and positive development that will bring needed accountability, transparency and public oversight to the water crisis as conflicts over the water Commons loom on the horizon(Maude Barlow). The problem of pollution due to solid wastes in Plachimada came to international attention through the BBC report on 25 July 2003. BBC reported the presence of heavy metals, lead and cadmium, in quantities higher than the approved limit in the sludge supplied by the Company, which was at the time dumping the sludge was a fertilizer. The heavy dumping of the sludge in the agricultural fields had been reported by Jananeethi, a non-governmental organization in Kerala in 2002 (Sujith Koonan, 2007-05).

An expert panel (A thirteen member High Power Committee) was formed in September 2009 and was supposed to submit its report by December 31st 2009. It had held four public hearings, including one at the Perumatti panchayat within the jurisdiction of which the beverages unit was located, and visited the affected areas for an on-the spot study of the damage allegedly caused by the Coca-Cola unit. It stated in March 2010 that the Company was accountable for damages to the village economy and its environment. The Company should pay adequate compensation to the affected people for the agricultural losses, health problems and the pollution of water resources. The Committee also recommended for a Plachimada Claims Tribunal to be created by the a state legislature for the purpose of claiming the compensation, amounting to around Rs 2162.6 million, from the company and distributing it to the affected people. But the company discarded this ruling. The company has not accepted the decision to form the expert panel and had even questioned its authority to go into the compensation question. The Plachimada Struggle Solidarity Committee and the Adivasi Samrakshana Sanghom, which were in the forefront of the Plachimada struggle, have been campaigning for extraction of compensation form the company and had submitted a detailed memorandum to the panel seeking compensation for the people of Plachimada and initiation of prosecution measures against the company (Gouridasan Nair, 2010).

> **Mailamma** (anti Cola woman leader) in her sixties with her four children had a long and arduous struggle. She belonged to a native tribal community, Eravalar. She was directly affected by Company's operations in Plachimada. She was a face that appeared at the satyagrahis, collecting everything from the public who supported their resist. She appeared at the satyagraha sight everyday; whether it was rain or sunny or she had job or not. She was the public face of the campaigning of the adivasi women who demanded that their water be resorted to them. This gender presence was of great significance, especially against the corporate hegemony. This aims at a reconfiguration of well-built power relations with micro effects that reach beyond the parochially defined and spatially situated notion of individuality. This she has maintained through democratic intervention, protest and arbitration. She played an important role in the campaign to hold Coca-Cola accountable for water shortages and pollution in the area. She died on January 6, 2007.

This being the existing situation in Plachimada, it remains to be seen how the struggle further goes mainly on the question of appropriation of water for substitute enterprises, the reinstatement of ecosystem, and compensation for damages caused. Private interest groups systematically ignore the option of community control over water. Because water falls on earth in a dispersed manner, because every living being needs water, decentralized management and democratic ownership are the only efficient, sustainable, and equitable systems for the sustenance of all. Beyond the state and the market lies the power of community participation. Beyond bureaucracies and corporate power lies the promise of water democracy. Water is a commons because it is the ecological basis of all life and because its sustainability and equitable allocation depend on cooperation among community members. Although water has been managed as a commons throughout human history and across diverse cultures, and although most communities manage water resources as common property or have access to water as a commonly shared public good, privatization of water resources is gaining momentum (Shiva, 2002).

Conclusion

Beginning with a struggle for livelihood security due to the operations of the Cola Company, the people of Plachimada laid the foundation of a historic struggle form below. It was from their every day experiences that they proved that their very lives were being put in danger by the activities of the multinational, resulting in, and their livelihood insecurity. The most striking of the struggle is the participation of women, who have not let a single day pass without their presence in the thatched shed in front of the Company. This gender power, led by Mayilamma was of great significance. By successfully resisting the advance of corporate hegemony through democratic mediation, protest, and negotiation, Plachimada has become a metaphor for a political village-society of contemporary times. In a broader context, the issues involved are not those that should be of concern to the village community alone, rather it falls within the remit of the larger resistance against Multinational power. The authentic value of a movement lies not just in addressing a wider democratic politics of worldwide connotation. This is particularly true in this context of water-related struggle, for at this point in time it is water in its multi-dimensionality that forms the object of the business gaze.

Suggestions

- India's National Water Policy should ensure that the water law not only achieve efficiency in the allocation of water resources, but also equity in their allocation and sustainability in the scale and mode of usage.
- India is yet to develop a well spelt out strategy on water privatisation and management of water resources. An urgent and early thought is necessary to certify a distinct framework on managing the process of water privatization.
- Build the custom of maintaining sustainably the community utilities. Long term neglect has always been the cause of failure.
- Encourage environmental conservation and protection, to facilitate sustainability of natural sources of water and

hence avoid advanced technology for water extraction that has always been used as a motive to encourage privatization.

- ☆ Promote the local means of harvesting and reserving water.

References

Bansil, P.C. 2004. *Water Management in India*, Concept Publishing Company, New Delhi.

Barlow, Maude, Our Water Commons: Towards a new freshwater narrative, The Coucil of Cnadians, www.onthecommons.org, Accessed on 12th Septermber 2010.

Charyula, Kumara *et al.,* 2007. Integrated Water Management-Hunger Free India, *Kurukshetra*, September.

Gathia, Sanjay. 2005. *Privatization of Water and Its Impact on the Right to Water in India*, MA Thesis, Mahidol University.

Gouridasan Nair, C. 2010. Coca Cola Must Pay Damages-Says Panel. *The Hindu*, Thiruvananthapuram Edition, March 15.

Jha, Nitish. 2010. *Access of the Poor to Water Supply and Sanitation in India: Salient Concepts, Issues and Cases*, Working Paper No.62, International Policy, Centre for Inclusive Growth, UNDP.

Kathpalia, G.N. and Rakesh Kapoor. 2004. Water Policy and Action Plan for India 2020: An Alternative, *India Vision 2020*, Report of the Committee on India Vision 2020, Planning Commission, Govt. of India, p.740.

Koonan, Sujith. 2007-05. Legal Implications of Plachimada: A Case Study, Working Paper, *International Environmental Law Research Centre*, Geneva, p.5.

Kothari, Ashish and Anuprita, Patel. 2006. *Environment and Human Rights*, National Human Rights Commission.

Raman, Ravi K. 2010. Transverse Solidarity: Water, Power, and Resistance. *Review of Radical Political Economics*, 42(251), pp.256-57.

Shiva, Vandana. 2002. *Third World Traveler, Water Wars*, http://www.thirdworldtraveler.com/Vandana_Shiva/Water_Wars_VShiva.htmlAccessed on 12th September 2010.

World Water Council. 2000. *'Ministerial Declaration of the Hague on Water Security in the 21st Century'*, The Hague, The Netherland, www.waternunc.com/gb/secwwf12.htm, accessed on 16th September 2010.

Wramner, Eva. 2004. *Fighting Cocacolonisation in Plachimada: water, softdrinks and a tragedy of the commons in an Indian village*, Lund University, Human Ecology Division.

Chapter 17

Waters of the Himalayas: The Natural Resource Fundamental to Life and Livelihoods

Vir Singh

Introduction

Origin of life on the planet Earth began with the availability of water. Water is not only essential, but fundamental to life and livelihoods, vital for survival and sustainability. Availability of water on sustained basis is a prerequisite for sustainable food security and food sovereignty. Water is not only needed for drinking, irrigation, generation of electricity, increasing land productivity, transportation, industries, aquaculture, home gardens, cleanliness, domestic use, etc., it serves virtually all purposes in life, directly or indirectly. The whole chemistry of life stems from water. Therefore, all debate of transformation of systems and socio-economic development, in essence, relies on the availability of water.

The superb issue, therefore, is that relating to water. There is plentiful of water on Earth: 71 percent, as we all know. The Earth is a water planet, perhaps the only water planet in the Cosmos. Water is also omnipresent in the biosphere, in one or the other form of the matter. It is water cycle that, in essence, maintains the very ecological integrity of the Earth's systems. This is the supreme role of water on Earth.

Impairment of water cycle with consequent serious disturbances in the regular flow of water that ensures sustained supplies of the resource to keep the life activities and development projects going on is the root cause of the problems. The superb issue relating to water is how to maintain water security, for this is imperative for the food and livelihood security and sovereignty everywhere on Earth.

With the rapidly growing water crises, competition over water among various uses and users is intensifying both within and between areas of a country and even between countries of the world. The water crises, in essence, emanate from the ecological and environmental crises. These crises have taken Himalayan mountains in their fold. Crises emerging from a macro ecosystem cannot be easily arrested or reversed. These expand rampantly from one area to another and from one region to another through ecosystem linkages.

Water crises arising out of the ecological degradation of mountain areas – due to vital impact of the vertical ecosystems interfering global environment – are more rampant and catastrophic than otherwise would be the case. In this Chapter we shall examine aspects relating to immaculate natural resource of water in the specific context of the Himalayan mountains of Uttarakhand.

The Himalayan Waters: The Third Pole on Earth

Most of the water sources on Earth have been polluted to a certain degree. Only the Himalayan waters (and some sources in the other mountains and highlands of the Earth) are regarded to be pure sources for potable water. The Hindu Kush-Himalayas (HKH) into the boundaries of the eight Asian nations – Afghanistan, Pakistan, India, China, Nepal, Bhutan, Bangladesh and Myanmar – are the largest storehouse of fresh water in the lower latitudes and as such serve as important 'water towers' for nearly 500 million people.

The highest, the youngest and the most fragile mountains of the HKH provide origin to major river systems, *viz.*, the Indus, the Ganga, the Yarlung-Tsangpo, the Brahmaputra, the Nu Salween, the Yangtze, and the Mekong. The HKH mountains are also called as the 'Third Pole', for they contain the largest mass of ice and snow outside the Earth's polar regions. There is a permanent snowline above 5000 m. Some of the glaciers in the Region are longest outside the poles of the Earth (Banskota and Chalise 2000).

To keep the 'Third Pole' of the Earth preserved through assured conservation of the water resource is one the greatest changes for our contemporary world. The HKH mountains are the common fragile natural resource. For the conservation of the waters of this Region, the entire world needs to extend help and support. As the mountain ecosystems have enormous bearing on the Earth's systems, their special care, regeneration and conservation of their pristine resources would fetch happiness, peace and prosperity to the larger parts of the world. The Agenda 21, Chapter 13 of the United Nations underlines the importance of the mountains for the world as a whole: "mountain environments are essential to the survival of global ecosystem".

The water resources in the Indian Himalayas are estimated to be about 1,20,000 million cubic meters in average annual flow (CSE 1991). Of the three major rivers in the Himalayan Region, the Brahmaputra, Ganga and Indus have an average annual flow of about 606.7, 390.0, and 210.0 million cubic meters per year, respectively (Valdiya, 1997). The Himalayan glaciers that cover about three million hectares of about 17 percent of the terrain are a unique reservoir that supports the mighty river systems. Controversy withstanding, the evidence shows that there is enough water in the Himalayan Region; a 'too-liitle-too-much-water' syndrome (Saravanan, 2000).

Magnitude in Uttarakhand

The Uttarakhand mountains are especially rich in their water resources. This area is home to dozens of perennial streams and numerous other rainfed rivers along with innumerable rivulets, waterfalls, and ponds, etc. India's largest river system – the Ganga River System – also takes its origin in the Uttarakhand Mountains. In Garhwal Himalayas, the largest glacier Gangotri, the origin of the river Ganga, has an estimated volume of well over 20 km^3 of ice, which is much more than even the maximum capacity of storage created by the Bhakra Dam on Sutlej, which is less than eight km (Sehgal, 1995). This glacier, unfortunately, now continues to recede. Apart from these streams and rivers, there are several lakes in the area as also the perpetual ice fields and glaciers.

However, due to maldevelopment, mismanagement, and continuous neglect, the Himalayas' aquatic resources are in hot waters. Dry spell all over can be witnessed especially in summer season. In the land of plentiful water, most of the villages are suffering from the lack of drinking water. Majority of the people do not have easy access to potable water. Only some 10 percent of the arable land is irrigated, and the rest 90 percent depends on rain. So gloomy is the situation in the area that drains almost the entire plains of North India (Singh, 2005).

Uttarakhand's glacier-fed rivers are being seen only as a potential source of hydropower. Mega projects carrying potential risk to ecology as well as to humanity, like the Tehri Dam Project, are being implemented. Ruthless mismanagement of Himalayan waters might trigger potentially negative consequences not only in Uttarakhand, but also in the entire Indo-Gangetic Valley. Aquatic ecology in the mountains, on the whole, is in a bad state, affecting the integrity of the entire food production system (Singh, 2005).

Integrated Management

All the water bodies in the mountains can be managed as a potential source of food production through aquaculture practices. The mountain water sources harbour a variety of food sources; fisheries and aquatic food plants being the most important. Water also has tremendous bearing on all components of the food production system.

Water, in fact, is the most important element in the ecological integrity of the farming system. Its very wise use and conservation-based management is vital for sustainable food security in the mountains. Water security and food security go hand in hand. Our efforts for ensuring water security must precede our efforts for food security.

For water security, fragile mountains obviously require high density of climax vegetation with enormous diversity of all types of plants – trees, shrubs, herbs, creepers, etc. Such vegetation, in addition to contributing biomass to cropland so crucial for soil fertility management, also responds to the problems associated with fragility of the Himalayan mountain ecosystems (Singh, 2005).

The extreme variability of climate and precipitation and patterns, as well as extremely inadequate knowledge on the hydrology of the mountain rivers and streams and the complex interrelationships between ecology and hydrology in the region impose serious scientific and technological limitations on the development of Himalayan waters (Banskota and Chalise, 2000).

While coverage of larger mountain slope areas with dense climax vegetation is the key measure to ensure conservation-oriented management of water resource, ecological farming practices with special care to biodiversity, water and soil conservation through contained soil erosion; minor irrigation systems built through local perspective planning; water harvesting techniques requiring prudent technology; and rational use of water resources to be made effective through policy perspective are some other important means of water management (conservation and utilization) in mountain areas.

We often hear many concerned citizens and environmentalists saying that the Third World War would be over water resources. Looking across the current intensifying conflicts between groups, states and countries, a larger war over water appears to be in the gestation period in the womb of the future. Water management, undoubtedly, is a key concern to world peace also. A radical change in national water policies with global consensus is also an obvious imperative of our times. Since the Himalayan area is shared by many countries in South Asia, a separate Himalayan Policy might be instrumental in restoring one of the most precious natural resources of the Earth.

References

Banskota, M. and Chalise, S.R. (eds). 2000. *Waters of Life: Perspectives of water Harvesting in the HKH*. Kathmandu: ICIMOD.

CSE. 1991. *Floods, Flood Plains and Environmental Myths: State of India's Environment*. A Citizens' Report 3. New Delhi: CSE.

Sarvanan, V.S. 2000. Policies, Programmes and Institutions in Promoting Local Water Harvesting in the Indian Himalayas. In Banskota, M. and Chalise, S.R. (eds.) *Waters of Life: Perspectives of water Harvesting in the HKH*. Kathmandu: ICIMOD, 29-49.

Sehgal, K.L. 1995. Aquatic Resources of North-Western Himalaya. In Singh, V.; Sharma, R.J.; and Kumar, A. (eds.) *Ecological Carnage in the Himalaya*, 35-47. Dehradun: IBD.

Singh, V. 2005. *Agrobiodiversity, Sustainability and Food Security in the Himalayan Mountains: An Uttarakhand Perspective*. Gorakhpur: GEAG.

Valdiya, K.S. 1997. Environmental Status Assessment: The Himalaya. In *Himalaya*, Vol. 18, No. 12. New Delhi.

Chapter 18

Effect of Natural Coagulants on Water Quality Parameters

L. Udaya Simha, Priya. V and Puttaswamy

Water is undoubtedly the most essential resource for man kind's survival. However, due to the recent developmental programmes worldwide, most of fresh water sources have been polluted in one way or other. The economics is one of the prime constraints in the water treatment unit processes. Thus the researchers have been investigating the usage of locally available material for the treatment of water and wastewaters. In this direction, the present investigation is planned to use the seed powder of natural bio materials like Drum Stick (Moringa Olifeira, M.O). Soybean (Glycine max, G.M), Pea (Pisum sativum, P.S), and Ground nut (Arachis Hypogea, A.H.) for the coagulation process. The results of these are compared with the most widely adopted chemical coagulant, alum. The changes in the quality of basic parameters of the water are also studied before and after coagulation process to assess the usage of bio materials as coagulants. The results revealed that they are found be efficient in reducing the turbidity to the order of 95-99 per cent, which is on par with the alum. But the water quality parameters, BOD and COD are observed to be increased by 2 to 3 times the raw water value. An attempt is also made to use alum as coagulant aid in the ratio of 80 per cent natural coagulant and 20 per cent alum. The results revealed that apart from reducing the turbidity to the order of 100 per cent, it is also observed the reduction in BOD and COD of the water sample. Thus the combination is proved to be beneficiary in all aspects and it can be opined that the natural seed powder can be used as coagulants in combination with alum as a coagulant aid.

Introduction

Water is the one of the basic necessities for humans. It is very well known that the available fresh water is limited, and in the present world, the same being polluted to an extent that the available water is still diminishing with respect to quality and quantity. Under these circumstances to supply potable for a community is not an easy task. The economics is also one of the prime constraints in the treatment unit processes. Thus the researchers have been investigating the usage of locally available material for the treatment of water and wastewaters.

Coagulation is one of the important processes in water treatment. It is an effective method in removing natural organic matters (NOM) and colloidal particles in high and intermediate molecular weight ranges (Sinsaubaugh *et. al.*, 1986). This process has a direct impact on the reliability of plant operations and final water quality and has significant contribution to the operational cost of the treatment plant (Hooge, 2000).

Aluminium salts are widely used as a chemical coagulant in water purification process all over the world. However recent studies have raised doubts about the advisability of introducing aluminium into the environment, especially concerning about residuals in the treated water, large production of sludge volume and Alzheimer's disease(Ndabigengesere *et al*, 1998; Okuda *et al*,1999). Many investigators have raised the doubts about the advisability of introducing aluminium into the environment by the continuous use of aluminium sulphate as a coagulant in water treatment (Letterman and Driscoll, 1988; Miller *et al.*, 1984). The research findings of Crapper *et al.* (1973) and Miller *et al.* (1984) showed that the chemicals used for water purification can cause serious health hazards if an error occurs in their administration during the treatment process. These reports suggested that a high level of aluminium in the brain is a risk factor for Alzheimer's disease.

Another problem of alums is reaction with natural alkalinity present in the water leading to the reduction in pH and low efficiency in coagulation in cold water (Katayon *et al.*, 2006). Ferric salts and synthetic polymers have been used as alternatives to aluminium salts. Due to high cost and low availability of such chemicals many developing countries can not afford to adopt this type treatments to water and waste water treatment (Diaz *et al.*, 1999; Schulz and Okun,1984).

To avoid the aluminium content in the environment, the researchers are attempting to find the alternative to alum. Natural coagulants of vegetable and mineral origin would be choice as they are used in water treatment before the advent of synthetic chemicals like aluminium and ferric salts. (Ndabigengesere *et al.*, 1998).

The research on natural coagulants conducted by the investigators (Jahn and Dirar, 1979; Jahn, 1984 and Sutherland *et al.*, 1989) financed through the German Agency for Technical Support and Oversees Development Administration, aimed at providing adequate water to rural people in Sudan reported that natural materials of plant origin can be used as coagulants. The mechanism of removal of turbidity is found to be from the seeds which can yield oil and has presence of protein in the seed (Wegelin and Metzler, 1994). The main parameter in reducing turbidity in natural coagulant seeds is the Protein. The protein is made up of amino acids, which are depicted as building blocks. Generally 19 to 20 kinds of proteins can be found in the seed protein. (Anhwange *et al.*, 2004) The efficiency of the removal of turbidity depends on the method of extraction of active agents from the plant origin.

Jahn (1979) investigated the potential of plant seeds and roots on removal of turbidity and compared it with alum. The researcher used clean water to extract the coagulating agent from seed powder and roots analogous to what is done traditionally by women in Sudan. Okuda *et al.* (1999) reported that M.O coagulant protein can be extracted by water or salt solutions (common NaCl).The amount and effectiveness of the coagulant from the salt and water extraction methods varies significantly. In crude form, salt extracts show better coagulation performance than the corresponding water extract. However purification of MOCP from the crude salt extract may not be technical and economically feasible. Marobhe (2004) has used distilled and spring water to extract active coagulating agent using very fine seed powder. The results revealed that, natural coagulants are most effective at very low concentrations that can make women in rural area to save up to 80 per cent of the seeds they use for clarifying water. Also the local coagulants have the potential to replace or supplement the synthetic organic polymer, which is imported at high foreign currency.

Apart from reducing the turbidity, the water quality is also an important criterion. Anselme N dabigengesere and K.Subba Narasaiah (1995) investigated on the quality of water treated by M.O. The measurements of pH, conductivity, alkalinity, Cation and anion concentrations showed that coagulation with M.O seeds did not significantly affect the quality of the treated water. However, concentration of organic matter in the treated water increased considerably with the dosage of Moringa Solution. Anselme Ndabigengesere *et al.* (1998), Maria Bergman *et al.* (2008) noticed that "Although the results of turbidity removal by the natural coagulants are efficient, but there is an increase in the COD due to the presence of Organic material."

Emelie Arnoldsson *et al.* (2008) conducted a study to compare between M.O. and Aluminium sulphate for coagulation of turbid water using jar test. The results revealed the coagulation with Aluminium Sulphate led to more efficient treatment but M.O. could also produce water of acceptable quality. Highest removal efficiencies are obtained when M.O. was extracted using tap water as compared to the distilled water and oil extraction. Treatment with M.O. did not change the chemistry of treated water and was more efficient for high initial turbidities. Francis Kweku Amalgloh and Amos Benang, (2009) carried out the investigation to confirm the effectiveness of powder extracted from mature dried Moringa Oliefera seeds. Findings of this research lent support to earlier works recommending the use of Moringa for water treatment.

Previous studies have shown that the seeds help in reducing the turbidity efficiently. But very few investigators studied the effect of the natural coagulant on the water quality. In the present investigation, an attempt is made to investigate the adoptability of four natural coagulants, *viz., Moringa oleifera* (M.O.), *Glycine max* (G.M), *Pisum sativum* (P.S.) and *Arachis hypogaea* (A.H.) as coagulants in comparison with a chemical coagulant alum, duly considering the quality parameters also.

Materials and Methodology

As indicated above four natural materials, *i.e.*, seeds of *Moringa oleifera* (drumstick), *Glycine max* (soybean), *Pisum sativum* (pea) and *Arachis hypogaea* (ground nut) and chemical alum have been considered for the study.

The coagulation process with the natural coagulants is reported to depend on the active protein present in the extracted solution (Ndabigengesere, 1998; Zhang, 2006; Davis, 2006). The literature indicates that the better efficiency is observed even when tap or distilled water is considered for the extraction process. In the present methodology also distilled water is considered for the extraction processes.

The preparation of coagulants from the natural seed materials is carried out as follows. The seeds are dried for a day in an oven at 40°C. The dried seeds are then ground in a kitchen blender and sieved through an iron sieve to yield a fine powder (200 micro meter). Ten grams of resulting natural coagulant powder is blended for 10 minutes with a kitchen blender. This blended mixture is made into 1 liter through the addition of distilled water, which gave a coagulant strength of 10 mg/ml (Mbogo, Shaaban Aman, 2008). The resulting extract is used throughout this study.

The water sample is collected from the Darga lake at Hosur, which is about 40 km east to the Bangalore city. The Physical and chemical parameters of the Raw and the Clarified water are determined according to the standard methods (APHA, 1992). This included the test for temperature, Turbidity, alkalinity, hardness, pH, total solids, total dissolved solids, total suspended solids, DO, BOD and COD.

Experimental Investigations and Analysis

The basic physio-chemical characteristics, as mentioned earlier, of the collected raw water are investigated before coagulation process. The average values of the raw water is presented in Table 18.1.

The seed extracts of four natural coagulants namely, Moringa oleifera, Pisum Sativum, Glycine Max, Arachis Hypogea are investigated with the removal of turbidity, as one of the objectives. To investigate the same, the optimum dosage of the natural coagulants are obtained by the Jar tests conducted at the laboratory.

The removal efficiency of natural coagulants is compared with the alum. The results are tabulated in Table 18.2.

Table 18.1: Characteristics of Raw Water

Charecteristics	*Values*
Turbidity	393NTU
Alkalinity	218 mg/l $CaCO_3$
Suspended solids	240 mg
Total Dissolved solids	10 mg
Hardness	136.1 mg/l $CaCO_3$
Temperature	25ºC
BOD	35 mg/l
COD	120 mg/l
pH	7.7

Table 18.2: Turbidity Removal by Natural Coagulants and Alum

Treatment Dose (mg/l)	*Residual NTUs in Water after Treatment*				
	Moringa oleifera	*Glycine max*	*Arachis hypogea*	*Pisum sativum*	*Alum*
0	393	393	393	393	393
100	208	208.8	218	209	173
200	24.3	24.6	28.5	16.3	21.1
400	6.23	18.2	23.4	10.9	10.6
500	4.3	16.3	19.7	7.5	7.4
600	5.1	12.4	23.15	6.9	4.2
800	5.2	11.6	26.3	5.2	3.6
1000	xxx	10.42	xxx	4.3	0.9
1200	xxx	8.13	xxx	4.5	1.2
1300	xxx	8.26	xxx	4.9	1.2
1400	xxx	8.31	xxx	xxx	xxx
% Turbidity Removal	99	98	96	99	Almost 100

From the Table 18.2, it can be observed that among the natural coagulants investigated, M.O. is found to be most efficient. The optimum dosage of M.O. is found to be occurring at 500 mg/l with the efficiency of removal 99 per cent, followed by P.S. (99 per cent at 1000mg/l), G.M. (98 per cent at 1200mg/l), A.H (96 per cent at 500mg/l) and for the chemical coagulant, alum, the optimum dosage is found to occur at 1000mg/l (almost 100 per cent). However, the optimum dosage of M.O. is half of that of alum and would be economical compared to the alum.

Turbidity removal according to the WHO Guidelines for Aesthetic Quality is 5NTU (Sinsaubaugh *et al.*, 1986). But the turbidity reduction depends on the initial concentration. In the present investigation for the initial concentration 393 mg/l, it can be observed that M.O. and P.S. are satisfying the norms. Whereas for GM the value is slightly above and for AH it is quite high. Hence the order of preference after M.O., with the present material used, would be P.S. and G.M. in satisfying the WHO standard.

The natural materials considered are found to be efficient in reducing turbidity to the order of 96-99 per cent. But the adoptability of the natural coagulants is also based on the quality requirements of treated water. The parameters considered for the investigation are pH, alkalinity, hardness BOD, COD, suspended solids, total dissolved solids. These parameters are investigated at the optimum dosage levels of natural coagulants and alum. Table 18.3 shows the comparison of the parameters with the source, *i.e.* raw water (R.W.) quality, treated water with natural coagulants and alum.

Table 18.3: Comparison of the Parameters with the Raw Water Quality

Substance	*Temp*	*Turbidity*	*pH*	*Hardness*	*Alkalinity*	*BOD*	*COD*	*SS*	*TDS*
R.W	25	393	7.7	136.1	218	35	120	240	10
R.W +M.O	26	4.3	7.7	144	196	70	240	128	352
R.W +G.M	25	8.13	7.6	140	200	110	380	206	453
R.W +P.S	25	4.3	7.7	140	196	103	440	152	650
R.W +A.H	25	19.7	7.7	141	198	106	320	192	530
R.W +Alum	25	0.9	6.8	135	132	22	210	21	632

It can be observed from the Table 18.3 that pH, hardness, alkalinity and temperature remained almost the same or varied very little with respect to raw water with the natural coagulants. The pH and alkalinity is observed to be decreased and hardness and temperature remained same with respect to raw water with alum as coagulant.

There is a drastic increase in the organic loading is observed as the natural coagulants are adding organic matter to the water sample. The increase in BOD with MO is found to be doubled and with other natural coagulants it is almost three times that of the raw water. Alum is found to efficient as it reduces the BOD by 37 per cent. The COD value of Raw water after treatment with natural coagulant is observed to be increased by three to four times its value. This could be due to the organic and inorganic constituents in the organic loading by the addition of natural coagulants to the raw water. With the alum also the same trend is observed. But the order of increase is less when it is compared with the natural coagulants. This could be due to alum precipitate out the organic constituents reducing the BOD of the raw water and hence the COD.

Variation of Suspended solids is observed to be a decreased in the concentration indicating all the four selected natural coagulants and a chemical coagulant are effective in decreasing the turbidity, the removal efficiency varies from to 15 per cent (GM) to about 99 per cent (alum) in the order G.M., A.H., P.S., M.O., Alum.TDS is found to increase in all cases supporting the increase in COD.

It can be noticed from the above investigations that the natural coagulants considered are increasing the BOD and COD levels in the treated water, So it can be opined that though the natural coagulants are efficient in removal of turbidity to 98-99 per cent, they can not be used as coagulants in their individual form since they are affecting the other water quality parameters. Many investigators (Sutherland *et al.*, 1996, Muyibi and Okuofu, 1995) have conducted the investigation with the combination of M.O. and Alum and found that this combination also efficient in reducing the concentration of turbidity. In the literature, the maximum efficiency is obtained by using 80 per cent of M.O. + 20 per cent Alum (Muyibi and Okuofu, 1995). Another set of experiments are planned and carried out using the same combination, *i.e.*, 80 per cent natural coagulants and 20 per cent alum. *Arachis hypogea* is not considered in this set as it has less efficiency in removal of turbidity when compared to the others and is not satisfying the WHO standards for the present experimental conditions. Below tables show the result of the combination.

Table 18.4: Optimum Dosage of the Combination 80 per cent Natural Coagulant + 20 per cent Alum

Treatment Dose mg/l	*Residual Turbidity in Water after Treatment, NTU*			
	G.M+Alum	*P.S+Alum*	*M.O+Alum*	*100% Alum*
0	393	393	393	393
200	24.3	9	6.3	21.1
400	16.5	8.5	1.7	10.6
500	10.7	4.6	0.5	7.3
600	3.1	2.4	0	4.2
800	0	0	0	3.6
1000	0	0	0	0.9
1200	0	0	0	1.2

From the above table it can seen that the optimum dosage for the combinations M.O+ alum, G.S.+ alum and P.S.+ alum are obtained at 600 mg/l, 800 mg/l and 800 mg/l respectively with turbidity removal 100 per cent.

The above experimental investigation has shown the positive result as per the turbidity removal. It has become necessary to investigate the other water quality parameters to study the feasibility of the combination at the optimum dosage.

Table 18.5: Characteristics of Treated Water with the Combination of Natural Coagulants (80 per cent) and Alum (20 per cent)

Characteristics	*Temp*	*Turbidity*	*pH*	*Hardness*	*Alkalinity*	*BOD*	*COD*	*SS*	*TDS*
G.M+Alum	26	0	6.7	132	140	13	117	12	445
M.O+Alum	25	0	7.1	142	163	8	93	6	353
P.S+Alum	25	0	6.6	130	137	13	103	19	258
Alum	25	1.2	6.8	135	132	22	110	21	632
Raw water	25	393	7.7	136.1	218	35	120	240	10

It can be seen from the table that the pH, alkalinity are decreased in all the combinations. This variation could be due to the presence of alum. However, the value of pH is near to the neutral and hence can be acceptable. Not much variation in the values of Hardness is observed in all the combinations. Most interestingly, the values of pollution indicators, *viz.*, BOD and COD are observed to be reduced using the combination. This could be related to the precipitation of the organic matter in presence of alum.

The above results show the combination is yielding the good results. It also can be noticed that though M.O is efficient, G.M and P.S are also equally effective. By adopting these natural coagulants in combination with alum, the quantity of aluminium in the environment can be reduced, which helps in sustainable development.

Conclusions

The natural coagulant materials considered for the study are found to be efficient in removal of turbidity to the level of 95-99 per cent. Among the four, the order of preference would be M.O. (99 per cent), P.S. (99 per cent), G.M. (98 per cent) and A.H. (95 per cent). The pollution indicators, BOD and COD are found to be increased by 2 to 3 times the original value with natural coagulants. COD is found to increase with alum also BOD is found to decrease with the alum. In their independent form, the adoptability of the natural coagulants may not be advisable. The combination, Natural coagulant (80 per cent) and Alum (20 per cent), is found to yield good result in reducing turbidity (100 per cent), BOD (65-70 per cent) and COD (3-16 per cent). The natural coagulant in combination with alum can be used in the coagulation process to attain sustainable development.

References

Anhwange.V.O., Ajibola and S.J. Oniye, Guill. 2004). Amino acid composition of the seeds of Moringa Oleifera(Lam), Detarium(Guill and Sperr) and *Bauhinia monandra* (Linn), Chemclass Journal, pp 9-13.

Anselme Ndabigengesere and K. Subba Narasiah. 1998. *Water Research*, Volume 32, issue 3, March 1998, pp 781-791.

Anselme Ndabigengesere, Narasaiah K.S., Talbot, B.G. 1995. Active agents and mechanism of coagulation of turbid waters using Moringa Oleifera, water research, 29(2):703-710.

Arnoldsson, Emelie and Bergman, M. 2007. Assessment of drinking water treatment using Moringa oleifera natural coagulant, Department of water resources, *Watten*, Vol 64, pp 137-150.

Crapper, D.R., Krishnan, S.S., Dalton, A.J. 1973. Brain Aluminum Distribution in Alzheimer's diesesse and Experimental Neurofibrillary Degeneration, Science 180, 511-513.

Davis, K.L. 2007. Alzheimer's Disease. Microsoft® Student 2007 [DVD]. Redmond, WA: Microsoft Corporation.

Diaz, A., Rincon, N., Escorihuela, A. Fernandez, N., Chacin, E., Forster, C.F. 1999. A preliminary evaluation of turbidity removal by natural coagulants indigenous to Venezuela. *Process Biochemistry*, **35**: 391-395.

Francis Kweku Amalgloh and Amos Benang. 2009. Effectiveness of Moringa oleifera seed as coagulant for water purification, African Journal of Agricultural Research Vol. 4 (2), pp. 119-123.

Hooge, P.A. 2000. *Evaluating Adsorption Clarification at Cold Water Temperature for Drinking Water Treatment*, University of Saskatchewan, M.Sc Thesis.

Jahn and Dirar. 1979. Studies on natural water coagulants in the sudan with special reference to Moringa oleifera seeds,Water SA, 5(2),pp 90,97

Jahn, S.A.A. 1984. Effectiveness of traditional flocculants as primary coagulants and coagulant aids for the treatment of tropical raw water with more than a thousand fold fluctuation in turbidity, water supply, 2(314), special subject, pp8-10

Katayon S, Noor M, Asma M, Ghani L. A. A, Thamer A. M, Azni I, Ahmad, Khor B. C. and Suleyman A. M. 2006. Effects of storage conditions of Moringa oleifera seeds on its performance in coagulation. *Bioresource technology*, 97(13): 1455-1460.

Letterman, R.D., Driscoll, C.T. 1988. Survey of residual aluminum in filtered water. *Journal of the American Water Works Association*, 80:154–158.

Maria Bergman *et al.*, 2008. Assessment of drinking water treatment using Moringa Oleifera, University essay from Londs University/Teknisk Vattenresuslara.

Marobhe. 2004. Natural coagulants and the extraction and purification methods of active coagulating agent: Reflection on applicability of methods in poor communities, Liter Course, 29 November, 2004. (www.lwr.kth.se/../Liter-course-NancyNov2006summary.doc)

Marobhe. 2004. Natural coagulants and the extraction and purification methods of active coagulating agent: Reflection on applicability of methods in poor communities, 29 November, 2004.

Mbogo,Shabaan Aman. 2008. A novel Technology to improve drinking water quality using natural treatment methods in rural Tanzania, Journal on environmental health.

Miller RG, Kopfler FC, Kelty KC, Stober JA, Ulmer NS. 1984, The Occurrence of Aluminum in Drinking Water, J Am Water Works Assoc, 76, 84-91.

Muyibi S. R., Okuofu. A. 1995. Coagulation of low turbidity surface waters with Moringa Oleifera seeds, International journal of environmental studies,1995,vol 48(3-4), pp 263-273, ISSN 0020-7233 CODENIJEVAW.

Ndabigengesere A, Narasaiah, K.S., Talbot, B.G. 1995. Active agents and mechanism of coagulation of turbid waters using *Moringa oleifera*. *Water Res*. 29(2): 703-710.

Ndabigengesere, Narasaiah K.S. 1998. Quality of Water Treated by Coagulation using Moringa Oleifera seeds, *Water Res*, 32, 781-791.

Okuda *et al.*, 1999. Coagulation Mechanism of Salt Solution-Extracted Active Component, *Water Research*, Vol 30, issue 11, june 2001, 2338-2344.

Okuda, T., Baes, A.U., Nishijima, W., Okada, M. 1999. Improvement of extraction method of coagulation active components from Moringa oleifera seed. *Water Res* **33**: 3373–3378.

Schultz, Okun D A (1984) Surface Water Treatment for Communities in Developing Countries, Intermediate Technology Publications.

Sinsaubaugh, R.L, R.C. Hoehn *et al.*, 1986. Precursor size and organic Halide formation rates in raw and coagulated surface water, *Journal of environmental Engineering*, 112(1): 139-153.

Sutherland, J.P., Folkard, G.K. and Grant, W.D. 1989. Seeds of Moringa species as naturally occurring flocculants, *Science, Technology and Development*, 1989, Vol.7, No.3, pp.1 91-197.

Sutherland, J.P., Folklard, G.K. and Grant, W.D. 1996. Natural coagulation for appropriate water treatment. A novel approach, *Waterlines*, 8(4):30 - 32.

Wegelin, M. and Metzler, A. 1994. A novel technology to improve drinking water quality using Natural coagulant, *Proc. Royal Society* 1893, 53:23-45.

Chapter 19

Climate Change, Ganga, Glaciers and Agriculture: Some Refletions

Girijesh Singh Mahra

Nature has sufficient to fulfill the need of all but nothing to satisfy the greed of one.

–Mahatama Gandhi

Once, I visited a hospital for my health checkup, There I saw a man with shattered skin, rashes all over the body; clear-cut it was the case of allergy. Doctors are rushing and making report on it. After a week I again visited that hospital for my scheduled checkup, Then I came to know a shocking news–that allergy case was the case of *Ganga bath*!!

Alas! Holy river Ganga, ancient river Ganga, that Ganga which takes away all the sins from mankind, today is causing allergy, irritation, skin rashes!!! That man has to pay price for his religious belief...

Also scientists got some samples from Ganga from specific places having disease-causing microbes. This small case itself indicates the change we have made in our nature. Change is the law of nature. Change is inevitable and change is an ever-going process, but *Climate Change* is one of the most complex, multifaceted and serious threats that the world faces today. Whether you are adding your bit to the heap of garbage piling up in your locality or not become meaningless when the garbage begins to rot, the stench will reach your nose too,

Climate Change induced by global warming works much the same way.

World famous scientists all over the country had evaluated climate change and came to conclusion that the Greenhouse Gases, CFC, Carbon dioxide and many other toxic gases are the cause of environmental degradation. Though it is true scientifically, the root lies elsewhere; that is in the greed of human being. Our greed led to advancement of technology and led us farther from peace and prosperity of all.

Increasing consumption of electricity due to initiation of several new industries, *disposal of garbage without treatment* to maximize profit, use of environmentally hazardous substances like polythene in our daily life are concrete examples of human greed.

People have been influencing the biosphere for at least 8000 years, since the invention of agriculture, but Climate Change has proposed a threat over our agriculture. The most affected area of Climate Change will be Agriculture and its biodiversity.

The first and the foremost impact of climate change is in biodiversity specially Islands biodiversity. nearly one –fourth of the world's countries are Island and they are treasure trove of biodiversity. Also they provide food, fresh water, wood, fibre, medicine, fuel and other raw materials.

But increasing sea level has given an alarm threat to them, The New Moore Island of India in Sunderbans has been consumed recently by rising sea. Many other Pacific atoll nation Island of Kiribati,Islands of Vanuatu were also submerged in early history. We are losing huge biodiversity; these are initial cases.

Rather IPCC has warned that a rise in sea levels of between 18 and 59cm by 2100 would be enough to submerge many other big islands, including Maldives and make them uninhabitable. Also of the 724 recorded animal extinctions in 400 years about half were Island species. Also climate change will bring degradation of coastal environment and natural resources on which poor rural people depend. Higher rates of erosion and coastal land loss may vanish our islands

In context of agriculture, sea level rise will also cause increased salinity due to encroachment of the sea and saltwater intrusion into freshwater lenses, contributing to an increasing shortage of water supply and loss of agriculture land. The most vulnerable section will be the poor and marginal farmers with small landholdings because *'extreme weather events'* will specially occur in tropics, fundamental changes in rainfall pattern together with rising temperature will shorten growing season and reduce crop productivity. According to the World Bank, 75 per cent of 1.2 billion people are trapped in extreme poverty. So they will fail to adopt new rotation practices and thus most vulnerable to climate change.

Worldwide farming is a significant source of greenhouse gas emissions accounting for 20 per cent of total emissions. Atmospheric concentration of methane has more than doubled during the past 200 years. Natural wetlands, fossil fuels related to natural gas, coal mines, coal industry, electric fermentation, rice fields, biomass burning, landfills account for 75 per cent of methane production in our environment. Thus major source of greenhouse gases is agriculture itself.

Rice fields are the most significant contributors of atmospheric methane accounting for 11-13 per cent of the world's total methane production. Also, rice production will need to expand by around 70 per cent over the next 25 years to meet the demands of fast growing human population whose food demand is expected to double in developing world in next 40 years. Thus, if we increase our crop area, especially the staple crop area in near future, then the problem of global warming is likely to increase.

Already in today's era 800 million people are unnourished, in India only 30 million people suffer hunger,46 per cent children are underweight and 17000 people are dying per day due to hunger on an average. Thus, in near future we are going to face a great threat not only of climate change but also of food security.

However, Not all effects of climate change on agriculture are expected to be negative. Most of the major food crops are C-3 plants including staple food like rice,wheat, oat, barley will show increase in yield ranging from 25 per cent to 64 per cent due to increased CO_2 level, also most of the noxious weed are C4 plants and their growth will be checked.

But this is only one side of the coin, increased temperature due to increase in CO_2 level may rise the pest attack up to 25 per cent and fungal disease attack up to 20 per cent and also disturb our monsoon pattern. In an overview it can be said that it will create more and new problem rather than benefits because the worst sufferers would be farmers of rainfed agriculture which covers 60 per cent of all cultivable lands.

Today, human beings have become dependent on technologies which are dependent on non -renewable resources and produce ill-effects in long run. Nowadays we say that glaciers are shrinking; may be shrinking but more that the heart of the people and their philosophy is shrinking too. While the drastic effects of global warming has been discussed in many scientific circles, had their run on T.V. and even had well meaning Hollywood films but the need of the hour is to change ourselves, to reduce our luxurious wants.

The need of the hour is to develop:

1. True sustainability: there should be synergies between climate mitigation strategies and development policies in areas of energy efficiency, fuel substitution, renewable energy, afforestration, and land and waste management.
2. The planning should not be for temporary economic gains and support, but for sustainability in future as well. There should be partnerships with communities, individuals, and private sector to frame effective measures to reduce the impact of climate change.

Today we all peoples of the world need to change our habits, not only in agricultural practices but also in our daily life practices.

Switching of fans and light before leaving room, efficient use of water at home, proper garbage disposal at micro level, though seems to be a very minute step but the huge building of true sustainability will only form above these basic steps because this will be the 1st step against our greed. So step ahead and realize your own responsibility, because it is not the time to curse the darkness but to light a small lamp.

References

Global Climate Change by Arnold J. Bloom

IPCC website

Know Climate Change by Tanya Agarwal

Science Reporter (CSIR)

Chapter 20

Treatment of Water Receiving Pulp and Paper Mill Effluents: Applying a Cost Effective and Affordable Method

Akanksha Rastogi and Vir Singh

Introduction

The Ministry of Environment and Forest, Government of India, has categorized the pulp and paper industry as one of the most polluting industries. (MOEF, 1998). Himalayan rivers and streams are especially of great ecological, social, cultural and economic importance. One of very specific roles of the fragile mountains of the Himalayas is reflected in the fact that they provide origin to a very large number of rivulets, rivers and streams, both glacier-fed and rain-fed. One of the largest river systems of the world – the Ganga River System – also finds its existence owing to the Himalayan mountains. These rivers are not only of critical importance for the Himalayan mountains, but also for the forelands in the plains. Healthy, clean and unceasingly flowing rivers are the best indicators of the ecological well-being of the mountains as well as of the plains.

Our rivers, unfortunately, are getting increasingly polluted. This is a matter of great environmental concern. Polluted rivers are slowly poisoning the land, livestock and human populations. Ill-health of rivers amounts to ill-health of people. Continuous disposal of human excreta and industrial effluents directly into rivers and streams is the major cause of water pollution of today.

Impact of cities on the environment is increasingly dominating the debate on sustainability. It is now evident that most global and regional environmental problems originate in cities. Cities accommodate and concentrate increasing number of people and human activities. Thus, they import increasing amounts of natural resources and export vast quantities of emissions and waste. Urbanization also entails major changes in the way people use natural resources, while it accelerates the transition from traditional to modern fuels. It also intensifies the use of energy and its environmental impacts. Economic development, increased industrialization, rapidly growing population and rising consumption of natural resources across the world are all impacting the environment significantly.

Pollution of environment is an undesirable change in the physical, chemical and biological characteristics of air, water and soil due to addition to the environment of material or energy (heat, noise and radioactivity etc.) in quantities and at a rate which is harmful to the living organisms including human beings. Pollution from pulp and paper industry further escalates this problem. When the pollution load flows through a stream, it is bound to affect whole socio-economic order for the inhabitants on the embankments due to pollution of underground water, wiping out of aquatic species especially fish, and also because of water's becoming unfit for consumption by humans and livestock and for irrigation of crops.

There are several methods of water pollution mitigation. Most of the methods, however, are expensive enough and not affordable by common people. This Chapter attempts to describe a method that uses a common, inexpensive and safe chemical, the alum.

Materials and Methods

Water pollution mitigation measure was carried out by using the alum [$K_2SO_4.Al_2(SO_4)_3.24H_2O$]. Water sample was taken from a stream that receives effluents from the Century Pulp and Paper Mill located in Lalkuan of Nainital district in Uttarakhand. After receiving the polluting effluents, the stream passes through dozens of villages and after about 18 kilometer long journey it is drained into Gola river near Kichha township in Udham Singh Nagar district. The water samples were taken from near Shantipuri, about 5 km away from the source of pollution, in the first week of June. Immediately after being brought to lab, the pH of the samples was measured using pH meter. Samples were also analysed for BOD and COD, which are regarded as pertinent indicators of the quality of water. Water samples with pulp and paper mill effluent in it were divided into six flasks. Each flask contained 500 ml sample. A lump of alum purchased from local market was powdered before dissolving in samples. Alum weighed on electronic balance in 0.25, 0.5, 1.0, 2.5, 5.0 and 10 g fractions was poured into six different flasks.

Each flask was vigorously shaken for two minutes for ensuring thorough mixing of alum and was left for four hours until the precipitation was clearly observed. The flask containing 0.25g in 500ml sample was left for about 10 hours to allow complete precipitation. Colour of each treated sample was noted down and the samples were analysed for pH, BOD and COD.

The pH of water sample was measured by an electronic pH meter. BOD was measured by Azide Modification of Iodometric method as described in APHA (1995). COD was measured according to the method described in Standard Methods for the Examination of Water and Wastewater (APHA, 1995).

Results and Discussion

The pulping process in a pulp and paper mill generates a considerable amount of wastewater which is highly polluting (Thomson *et al.*, 2001). Each pulping process utilizes large amount of water, which reappears in the form of an effluent. The most significant sources of pollution among various process stages are wood preparation, pulping, pulp washing, screening, washing, bleaching and paper machine and coating operation. Wastewater, which is coming from these processes, contains wood debris and soluble wood material. Wood consists of various compounds (lignin, carbohydrates and extractives) which are hard to biodegrade and these derivatives are washed away from the fibers during the washing, dewatering and screening processes (Katja and Mika, 2007).

The wood pulping and production of the paper products generate a considerable amount of pollutants characterized by BOD, COD, TSS, toxicity and colour when untreated or poorly treated effluents are discharged to receiving water (Katja and Mika, 2007). Since the effluents cause a number of environmental problems, like slime growth, thermal impact, scum formation, etc., they also increase amount of toxic substances in the water causing death to the aquatic plants and animals, severely affecting the terrestrial ecosystem. Therefore, effluents must be treated before discharging them into the environment.

Mitigation of the polluted water taken during the month of June from the Lalkuan-based pulp and paper mill drainage channel in Shantipuri about 5 km away from the location of the factory was carried out using different amount of alum. This chemical method was opted for the wastewater pollution mitigation because of the following reasons, *viz.*:

1. Low cost and easy availability of the chemical;
2. A natural germicide and coagulant or flocculent;
3. Requirement of lower amount of the chemical; and
4. A reliable chemical frequently used in average families as preservative and water purifier.

Alum is a salt that in chemistry is a combination of an alkali metal, such as sodium, potassium, or ammonium and a trivalent metal, such as aluminium, iron, or chromium. The most common form, potassium aluminum sulfate, or potash alum, is one form that has been used in food processing. Another, sodium aluminum sulfate, is an ingredient in commercially produced baking powder. The potassium-based alum has been used to produce crisp cucumber and watermelon-rind pickles as well as maraschino cherries, where the aluminum ions strengthen the fruits' cell-wall pectin. Alum is approved by the U.S. Food and Drug Administration as a food additive, but in large quantities – well, an ounce or more – it is toxic to humans (http://www.ochef.com/1080.htm).

Flocculants are used in water treatment. The addition of flocculants to raw water causes colloids and other suspended particles to stick together and form heavier particles (floc) which will be removed by the sedimentation or filtration. This flocculation (or coagulation) process is to aid the removal of contaminants like fine solid pollutants or microscopic molecules which are difficult or impossible to be removed by filtration alone. Generally flocculants are multivalent cations such as aluminium, iron, calcium or magnesium. Many of the suspended water particles have a negative electrical charge which repels each other. Positively charged flocculants attract and stick to many of the suspended water particles. Many of flocculant cations, under appropriate pH and other conditions, react with water to form insoluble hydroxides which join together to form larger settleable particles or physically trap small particles into the larger floc. There are organic flocculants also. Members of flocculants include alum, aluminium suphate, calcium oxide, iron chloride, iron sulphate, polyacrylamide, Polyaluminum chloride, Polyaluminum hydroxidechloride silicate sulfate, Polyaluminum hydroxidechloride silicate, Sodium aluminate, Sodium silicate (http://www.ochef.com/1080.htm).

Alum, one of the most common flocculants, was used to remove the pollutants from the wastewater coming out of the pulp and paper mill. Four indicators, namely, colour, pH, BOD and COD were used to see the level of pollutants in the samples. Values of these indicators in the wastewater and treated wastewater are presented in Table 20.1.

Table 20.1: Physico-chemical Properties of Wastewater Treated with Alum

Sample	*Amount of Alum (g/l)*	*Colour*	*pH*	*BOD*	*COD*
Control	0.00	Dark brown	7.94	120.00	300
1	0.50	Clear	7.69	90.00	298
2	1.00	Crystal clear	6.69	60.00	260
3	2.00	Clear	4.12	60.00	266
4	5.00	Light yellow, slightly translucent	3.84	56.00	269
5	10.00	Pale yellow, slightly translucent	3.73	54.00	276
6	20.00	Pale yellow, clear	3.64	54.00	302

The colour of the wastewater coming out of factory was dark brown. This colour is unacceptable in the water bodies and is the first conspicuous indicator of water pollution. Treatment with varying amounts of alum led to almost complete decolouration of the wastewater.

Application of just 0.50g alum per litre (or 0.05 per cent alum w/v) of wastewater turned the sample clear by decolouration.

Treatment by dissolving 1.0g alum per litre (or 0.1 per cent w/v) of the wastewater turned the solution crystal clear. The supernatant in appearance looked like drinking water. Again, increased amount of alum beyond 0.2 per cent w/v, imparts light yellow and pale yellow colour to the treated wastewater (Table 20.1).

The pH value of the treated water decreased with increased concentration of the alum. pH of control was 7.94 but that of the treated water with 2 per cent alum (20g per litre of water) was as low as 3.64. pH of the water treated with 0.05 and 0.1 per cent alum (w/v) was nearly neutral. Increase in alum concentration brought down the pH value up to unacceptable limit. Alum is acidic in nature and forms acidic solution with water (Table 20.1; Figure 20.1).

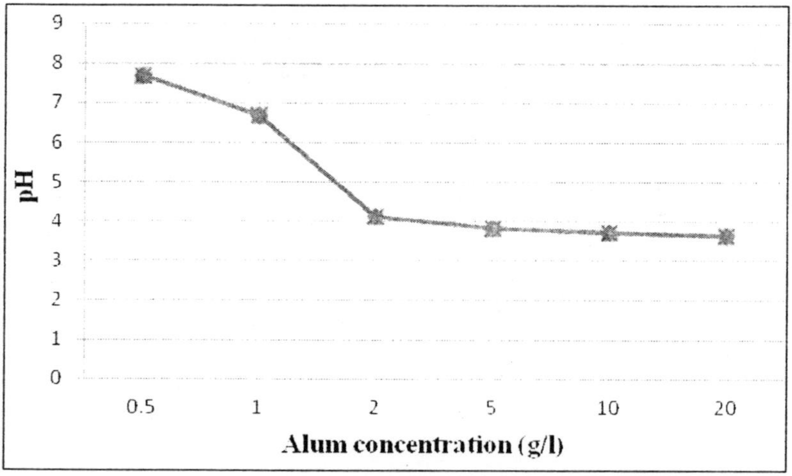

Figure 20.1: pH of Treated Wastewater with Varying Alum concentration

Considerable decline in the BOD of water treated with increased amount of water was recorded. There was 50 per cent decline in the value of BOD with dissolving just 1.0g alum per litre of water (Table 20.1; Figure 20.2). Alum acts as an anti-microbial chemical. It would suppress the growth of microbes in water. Demand of oxygen by microorganisms to decompose the organic matter would therefore be decreased leading to the lowering of BOD.

COD, however, is not brought down considerably. COD of the control and that of the waste water treated with 2.0 per cent alum

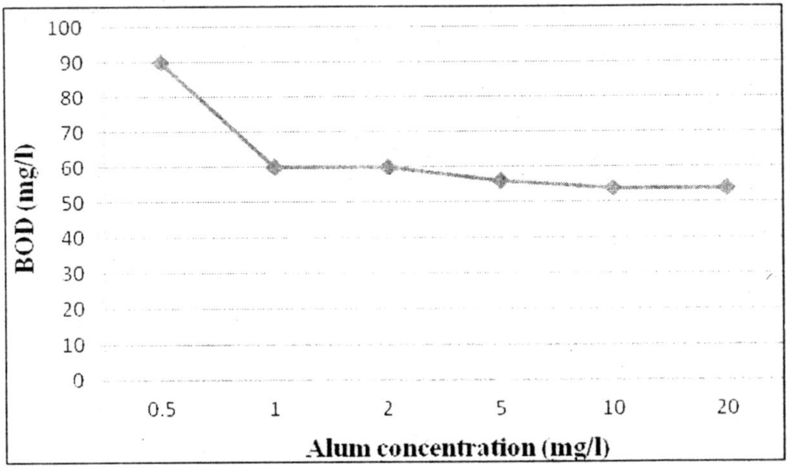

Figure 20.2: BOD of Treated Wastewater with Varying Alum Concentration

was almost equal. COD of the water treated with 0.1 per cent solution (w/v) was recorded to be the lowest (Table 20.1, Figure 20.3). Since alum is an inorganic compound, it adds to the inorganic load of water. Therefore, COD increased with the increase of alum concentration in the water beyona a certain limit.

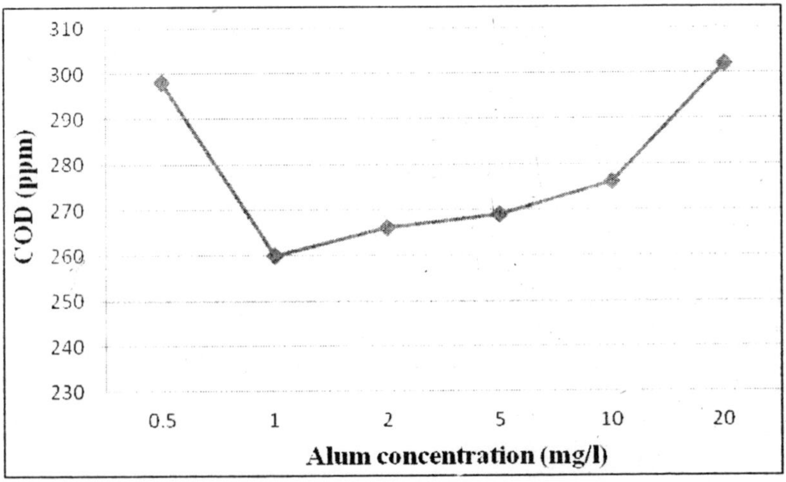

Figure 20.3: BOD of Treated Wastewater with varying Alum Concentration

Based on the physical, chemical and biological indicators, it could be concluded that wastewater purified itself to certain extent as it flowed further in the stream, except for the colour of the wastewater, which did not improve at all. However, the purification process did not go on in a linear fashion. The process is affected by a mix of various factors.

Mitigation of the polluted water was carried out using different amounts of alum. This chemical method was opted for the pollution mitigation because of low cost and easy availability of the chemical. Further, the alum is a natural germicide, an effective coagulant and flocculent and it is a reliable chemical frequently used in average rural Indian families as a preservative and water purifier.

The colour of the wastewater coming out of factory was dark brown. This colour is unacceptable in the water bodies and is the first conspicuous indicator of water pollution. Treatment with varying amounts of alum led to almost complete decolouration of the wastewater. Application of just 0.50g alum per litre (or 0.05 per cent alum w/v) of wastewater turned the sample completely clear by decolouration. The pH value of the treated water decreased with increased concentration of the alum. The pH of the control was 7.94 but that of the treated water with 2 per cent alum (20g per litre of wastewater) was as low as 3.64. The pH of the water treated with 0.05 and 0.1 per cent alum (w/v) was nearly neutral. Increase in alum concentration brought down the pH value up to unacceptable limit. Alum is acidic in nature and forms acidic solution with water.

Considerable decline in the BOD of water treated with increased amount of water was recorded. There was 50 per cent decline in the value of BOD on dissolving just 1.0g alum per litre of water. Alum acts as an anti-microbial chemical. It would suppress the growth of microbes in water. Demand of oxygen by microorganisms to decompose the organic matter would therefore be decreased. COD, however, did not decrease as the alum is an inorganic compound and contributes to increase COD on increasing its amount in the solution.

It could be inferred that treatment of wastewater with 0.1 per cent alum solution (w/v) is the most appropriate measure for the mitigation of pollution of the stream water due to pulp and paper mill. It completely decolourises the wastewater, brings the pH to a desirable level. However, the BOD of the sample is above prescribed limits, which needs to be addressed. This low amount of alum would not act as a pollutant and is acceptable even for consumption (provided BOD is mitigated). Moreover, costs of this chemical available everywhere will be quite low and will hardly have bearing on the cost of paper and pulp production.

References

APHA (American Public Health Association). 1995. *Standard Method for the Examination of Water and Waste Water*. 17th Ed. Washington, DC.4: 4. 45-4.67.

http://www.ochef.com/1080.htm.

Katja, H. and Mika, S. 2007. Flocculation in paper and pulp mill sludge process. *Research J. of Chemistry and Environment*, 11 (3): 96-104.

MOEF (Ministry of Environment and Forests). 1998. *Compendium of Environment Statistics*. MOEF, New Delhi, pp 218.

Thomson, G.; Swain, J.; Kay, M. and Forster, C.F. 2001. The treatment of pulp and paper mill effluent: a review. *Bioresour. Technol.*, 77, 275-285.

Chapter 21

Water, Energy and Women for Sustainable Development and Achieving Millennium Development Goals in the Mountains

Vir Singh

Himalayan mountains owing to the largest source of snow outside the two poles of the Earth – the North Pole and the South Pole – are often referred to as the Third Pole. These ecosystems provide origin to thousands of rivers, streams, rivulets, springs, lakes, etc. Himalayas are also regarded as "Water Towers for the Mankind" providing water security in the entire South Asia. Himalayas also provide some of the largest water systems to the world. Yet a sizeable majority of the mountains has no direct access to water. Himalayas' richness in water hardly serves local populace as much as it does to the downstream populations. Water, in essence, remains one of the scarcest basic needs of people in the Himalayan mountains.

Himalayan rivers are being used as one of the major sources of energy–the hydro-electric energy regarded to be one of the cleanest forms of energy. Large Himalayan rivers have been dammed; thousands square kilometers of geographical area in the fertile valleys have been inundated. Millions of people have been displaced from the submergence areas. But local Himalayan people are crying for energy. Their own development suffers from an acute shortage of useful energy. Hydro-electricity generated in the mountains ensures industrial and urban areas' prosperity in the plains. However, fuel wood extracted from local forests remains the major source of energy for cooking food and warming houses for the local populace.

A very impressive network of roads has reduced inaccessibility of the mountains. A very efficient transport system is another feather in the cap, further overcoming inaccessibility in the mountains. All the consumer items virtually of all the famed markets of the world are now available almost everywhere in the mountains. Tourists from all over the world can be found rejoicing even in the interior most areas of difficult accessibility. Mountain women, however, have not witnessed a significant improvement in their life styles. They are still found working under difficult circumstances, working in the fields, walking with head-loads of fuel wood and water containers on treacherous footpaths. Roads, transport, availability of modern amenities, etc. have hardly contributed to improve the lot of local women. Leisure, recreation, healthy life, enjoyment, etc. still remain a dream for the local women.

The twin problems of water and energy mountain areas face are the most tedious constraints to mountains' socioeconomic development. While acute water and energy crises are often discussed, women related issues are often missed. The three crucial issues are intertwined together; if one is not resolved, the other won't also be resolved. The three piercing issues need to be resolved together, only then shall we choose the right path to sustainable mountain development.

The Third Millennium has in its folds some new hopes for humanity. As the 20th Century and 2nd Millennium came to a close and the 21st Century and the 3rd Millennium ushered in Humanity saw new dreams. Meetings, conferences, symposia were held the world over and new dreams and visions came to the fore. New pledges were made to fulfill promises for the world. The most important was the Millennium Summit organized by the United Nations in 2000. In the Millennium Summit Millennium Development Goals (MDGs) were set. As many as 189 countries adopted MDGs, India being one of them. This article reveals that the resolution of the issues relating to water, energy and women could be important to achieve MDGs. The deadline for realizing MDGs is 2015 which is not far off.

The Current Scenarios

Worldwide 1.1 billion people lack access to safe drinking water, while 2.0 billion people continue to rely on traditional biomass fuel for cooking and heating. Of the 1.3 billion people who live in poverty, 70 per cent are women. Women perform two-thirds of the world's work but earn one-tenth of the world's income (Sharma and Banskota, 2006).

South Asia, the region with the largest population density in the world and widespread poverty, has been identified by many as an area of quickly emerging water crises. Excepting the island areas, the South Asian region covers about 3.3 per cent of the terrestrial are and receives 6.8 per cent of the annual replenishable amount of water of the world. With this land and water resources, the region supports 21 per cent of the world's population (Bandyopadhyay, 2006). An overwhelming population of South Asia depends on firewood for cooking. Petroleum import costs these countries exorbitantly. Women's status is also worst in many South Asian countries. Lack of literacy and education, socio-economic inequality and several cultural taboos imposed on women compel half of the humanity in the region to live in miserable conditions.

Himalayan region is particularly sensitive to pronounced temperature rise, particularly at high altitudes. The Hindu Kush-Himalayan Region has a vast snow-covered area (SCA); the maximum area recorded from 2002 to 2010 was 1.79 million km^2 (42.9 per cent of the total land area). The permanent SCA during this period was 0.18 million km^2 (4.3 per cent of the total land area). The SCA shows strong inter-annual variation (Figure 21.1). The mean annual SCA ranged from 0.70 million km^2 in 2002 to 0.84 million km^2 in 2005 (Gurung *et al.*, 2011). A nine-year (2002-2010) analysis of SCA is shown in Table 20.1.

Table 20.1: Annual snow cover trends in the Hindu Kush-Himalayan (HKH) Region (per cent)

Region	*Trend*
HKH Region	-0.05 ± 1.32
Western HKH	+0.02 ± 1.36
Central HKH	-0.40 ± 1.86
Eastern HKH	+0.03 ± 1.55

Source: Gurung *et al.* (2011).

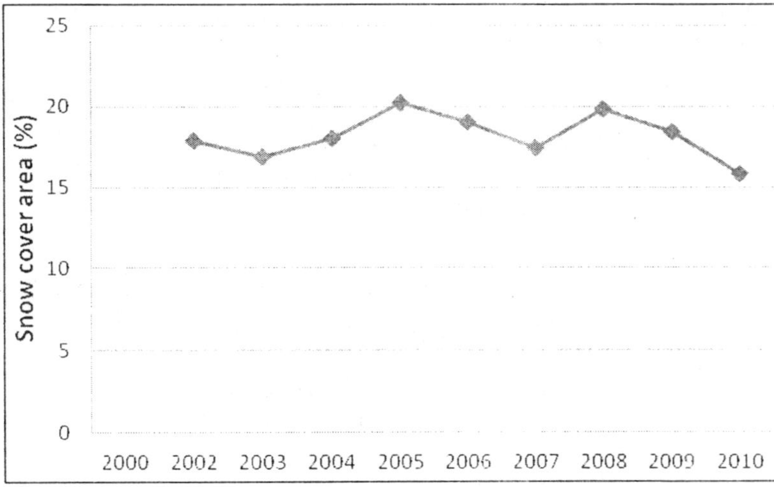

Figure 20.1: Average Variation in Snow Cover Area in the Hindu Kush-Himalayan Region, 2002-2010 (Based on data of Gurung *et al.*, 2011)

A multi-stage approach using both remote sensing and field data applied by ICIMOD revealed about 54,000 individual glaciers with a total area of 60,000 km^2 and an estimated 6100km^3 of ice reserves in the entire HKH Region (Table 20.2). The stored water in these ice reserves is about four times the annual precipitation. The highest concentration of glaciers is in the upper Indus basin, and the lowest in the Irrawaddy basin. The largest glacier in the HKH Region is the Siachen Glacier in the Indus basin (Gurung *et al.*, 2011).

The Himalayan areas receive heavy summer precipitation up to an altitude of about 2700 masl. Monsoon precipitation on the southern aspect of the Himalaya provides the most significant run-off in the region. The sediment load in Himalayan rivers poses serious challenges to water management in the Himalayan region. The largest of the Himalayan rivers, Ganga, Brahmaputra and Indus, are the international rivers. Growing use of Himalayan rivers for irrigation and hydro-electricity generation is causing serious environmental problems on one hand, and international disputes and conflicts on the other (Bandyopadhyay, 2006).

Institutional efforts are on way to develop alternative sources of energy in mountain areas. Harnessing solar energy to produce solar dryers, solar lanterns, and improved technologies to increase

Table 20.2: The Status of Glaciers in Hindu Kush-Himalayan Region

River Basin	*Number of Glaciers*	*Glacier Area (km^2)*	*Ice Reserves (km^3)*	*Average Glacier Area (km^3)*
Amu Darya	3,277	2,566	162.6	0.8
Indus	18,495	21,193	2696.1	1.2
Ganga	7,963	9,012	793.5	1.1
Brahmaputra	11,497	14,029	1,302.6	1.2
Irrawaddy	133	35	1.3	0.3
Salween	2,113	1,352	87.7	0.6
Mekong	482	235	10.7	0.5
Yangtze	1,661	1,660	121.4	1.0
Yellow	189	137	9.2	0.7
Tarim	1,091	2,310	378.6	2.1
Interior	7,351	7,535	563.1	1.0
HKH total	54,252	60,054	6,126.9	1.1

Source: Gurung *et al.* (2011).

fuel efficiency through 'improved *chulahs*' are often encountered. Liquid petroleum gas (LPG) distribution is also not rare these days, but the fact is this that rural women still suffer from various diseases due to cooking in suffocated environment of their kitchens. Head loads of fuel wood from distant forests consumes much of their time and keeps them under drudgery. In some of the HKH areas, such as in Bhutan, women's engagement as energy entrepreneurs under UNEP/ICIMOD projects (Sharma and Giri, 2006), however, are some interesting stories to derive lessons from.

Water-Energy Crisis and Implications for Women's Life

Accelerated glacier retreat in the Himalayan region is one of the most striking indicators of climate change. This phenomenon is being regarded as a grave threat to the mountains as well as to the foreland areas fed by perennial rivers originated from the Himalayan glaciers. Glacier-fed rivers in the Himalayan region are seen with reduced water flow during lean period. One of the largest river systems of the world – the Ganga River System -- would soon be in hot waters provided global warming trend continues to prolong progressively.

Many rainfed rivers with origin in natural forest areas often go dry a couple of months after monsoon. Mountains on Earth make rivers' origin and their flow towards oceans possible and thus impart a definite course to water cycle on Earth. Long routes of water flows through streams and rivers phenomenally contribute to continuous recharging of groundwater sources. But now ecological ruin inviting climate change appears to be submerging hopes that water promises to sustain.

Lucky are the countries which have hydro potential to generate electricity, for the hydro-electricity is not only least expensive to generate but also environmentally clean and helps avoid lot of pollution which would have not been the case provided the electricity were generated by coal or natural gas. The Himalayan mountains help harness the water potential for hydro-energy that translates into socioeconomic development. Lucky is India for it has Himalayas for paving way to hydro-electricity generation. But what Himalayan communities get in return? Only a meager amount of the energy produced within their ecosystems. This energy could be used in saving natural forests from going into fire-hearths for cooking food. This energy could be used in pumping drinking water from rivers to their houses in hilltops, in elevating water for bringing more of the cultivated area under irrigation cover, in running cottage industries in mountain villages, and in keeping hill houses warm during winter season. The energy generated in the mountains could altogether make a difference in people's lives. But only a small fraction of this energy goes to the benefit of mountain people. Overwhelming dependence on fuel-wood, continuous degradation of environment due to deforestation, continuous drudgery and stress on mountain women mostly due to carrying water and fuel-wood from long distances, are some of the negative indicators that tell the brutal story of mountain life.

Mountain areas in India are perhaps the most sought after targets for development intervention. Millions of Rupees are invested in the name of sustainable development of the Himalayas. Indian Council of Agricultural Research (ICAR) is operating Krishi Vigyan Kendras (KVKs) almost in all mountain districts. Agricultural universities, agricultural research institutes, Institute for Himalayan Environment and Development, numerous NGOs, government line departments etc. are all engaged in 'Himalayan development'. Tall claims of development are made in their reports. Their analyses bring feelings of "feel good". For whom has this so-called development taken place? Certainly not for mountain women! Their status has not improved at all. In many cases, women's status has worsened in actual sense. The roads, the transport, the tourism, the eco-tourism, the industries, the hotels, the resorts, the dams, the projects, the institutes, the universities are not for them. 'Drudgery of hill women' is often discussed in seminars, conferences, meetings, etc. but the state of common women's drudgery never changes. Development intervention is not gender-sensitive. It only favours the already-favoured groups–the male, the rich, the political elite. Mountain women appear to be the last in the preference of development agendas.

There appears to be a clear-cut disconnect within water, energy and women. Water and energy are essential resources not only for improving women's lot but also for human survival and well-being. Women are the life-line of mountain agriculture. A large proportion of animate energy in mountain agriculture flows through women's labour. In some cases women, in terms of energy output to agricultural operations, work more than bullocks (Singh, 1998). Scarcity of water and energy cause hardships to women and girls, who must spend many hours each day collecting and loading water and fuel woods from distant sources.

Water-Energy-Women and Millennium Development Goal

The Millennium Declaration adopted by 189 countries at the United Nations Millennium Summit in 2000 and the eight key goals (MDGs) to be achieved by 2015 have become the international roadmap for working on poverty eradication, health, education, and environmental sustainability (Sharma and Banskota, 2006). As there is a definite 'hidden' link between energy, water and women, none of the targets can be achieved without management of the two resources and women empowerment.

How the MDGs could be achieved through water-energy-women development processes is shown in Table 21.3.

Some Suggestions

The water and the energy, the key resources for socioeconomic, cultural and ecological development and sustainability, need be

Table 21.3: Linking MDGs with Energy-Water-Women Development Issues

Goal	*Target*	*Linking Water-Energy-Women Development with MDGs*
Goal 1: Eradicate extreme poverty and hunger	**Target 1:** Reduce by half the number of people living on less than a dollar per day **Target 2:** Reduce by half the number of people who suffer from hunger	☆ Water and Energy related technologies make women empowered enough to undertake new/ extra activities to improve their income. ☆ Newer farm technologies help women to increase food production and improve household income and nutrition of family members.
Goal 2: Achieve universal primary education	**Target 3:** Ensure that all boys and girls complete a full course of primary schooling	☆ Easy excess to water sources/ regular water supplies, availability of efficient fuels helps free children to spare time to go to school. ☆ Availability of energy sources for light at night helps children study at night. ☆ Energy-water availability leads to increased incomes to help pay school fees.
Goal 3: Promote gender equality and empower women	**Target 4:** Eliminate gender disparity in education	☆ Availability of water and energy on sustained basis help women participate in programmes like adult literacy and children going to school. ☆ Educated women are empowered women who play crucial role in achieving equality in society as well as in education system.
Goal 4: Reduce child mortality	**Target 5:** Reduce by two-thirds the mortality rate among children under five	☆ Clean energy (to eliminate indoor pollution) and safe drinking water reduce diseases and improve women's health. ☆ Women education helps to increase awareness to health, hygiene and sanitation issues.
Goal 5: Improve maternal health	**Target 6:** Reduce by three-quarters the maternal mortality ratio	☆ Freedom from excessive workload and drudgery due to carrying heavy loads of fuel wood and water from long distances has positive implications for women's health. ☆ Safe drinking water and cooking in smokeless environment improve women's health and well-being.

Contd...

Table 21.3–*Contd...*

Goal	*Target*	*Linking Water-Energy-Women Development with MDGs*
		✰ Education makes women aware of all health related issues and importance of balanced nutrition.
Goal 6: Combat HIV/ AIDS, malaria and other diseases	**Target 7:** Halt HIV/ AIDS, malaria and other diseases by 2015 and begin to reverse the spread of HIV/AIDS	✰ Public health is associate with women's awareness achievable through living in sane environment and through education, including sex education. ✰ Easy access to water and energy makes all the difference in women's life.
Goal 7: Ensure environmental sustainability	**Target 9:** Reverse loss of environmental resources **Target 10:** Reduce by half the number of people without sustainable access to safe drinking water	✰ Clean energy is the best bet for environmental conservation and human health ✰ Clean water supplies protect health and help keep environment clean ✰ Alternative sources of energy reduce/ eliminate pressure on forests to the conservation of lofty and fragile mountains and strike an ecological balance. ✰ Conservation of forests is critical for the conservation of water. ✰ Alternate energy sources substantially reduce carbon emission and forests saved through reduced biotic pressure enhance carbon sequestration–vital steps towards climate change mitigation. ✰ Educated and empowered women play critical role in conserving natural resources.

Source: Column 1 and 2 are based on Sharma and Banskota (2006).

managed in such a way that their supplies remain intact. Water is an inexhaustible resource. And there are forms of energy which are also inexhaustible. Integrated management of these resources would bring phenomenal changes to natural and human systems. Women empowerment is another issue vital for social development. Energy and water are closely linked with women empowerment and overall social development. Gender, of course, is a cross cutting issue virtually in all the important organizations of the world. But the situation is not in balance. Women are still staying on the margin. Solution of their problems is often sought without taking natural resources into development. Their empowerment is inevitably associated with the natural resources which are vital for social development. Water and energy are the two major resources which make difference to social life. The following focal points would be vital for changing scenario in the positive side:

- ☆ Integrated management of water and energy resources;
- ☆ Environmental conservation with focus on natural forests (their protection, conservation and augmentation) which play phenomenal role in conserving water, regulate water cycle and mitigate climate change necessary to halt accelerated melting and retreat of glaciers;
- ☆ Eco-hydrological dimensions of surface water, avoidance of damming rivers with big dams, promotion of small dams, run-of-the-river systems for energy generation from enormous water resources in the mountains;
- ☆ Balanced land use, with due proportion to natural forests, emphasis on the cultivation of landraces/local varieties of crops demanding less water, use of soil-water conservation measures in agriculture, avoidance of development intervention in Greater Himalayas;
- ☆ Development of mitigation measures to minimize impact of extreme hydrological events, such as droughts and floods;
- ☆ Understanding of social dimensions of water systems management and water conflicts;
- ☆ Greater emphasis in R&D on alternative sources of energy, especially solar energy for cooking, lighting, drying and warming houses;
- ☆ Supply of electricity to local populace at affordable rates;
- ☆ Development of women-friendly technologies aimed at eliminating drudgery from their normal routine activities;
- ☆ Ensuring higher participation of women folk in development activities right from planning to decision-making to implementation;

- Ensuring compulsory education for girls and promotion of women participation in higher education and a gender-balanced situation in the services of all private and public sectors;
- Ensuring environmental sanity, hygiene, primary health services;
- Constitution of effective local governance ensuring food and nutrition security, social and cultural equity, and equal access of all to natural resources, especially energy and water and health services.

Conclusion

Water, energy and women make an important axis for sustainable development of mountain areas. A gender-balanced development is sustainable development. Precluding women in socioeconomic, cultural and eco-development exercises would lead to chaos. Women's contribution to progress and happiness of this world is phenomenal. No hope of a sustainable future can sustain without their valuable contribution. Their own development, progress and happiness depend on vital natural resources, namely water and energy. This is especially true of fragile mountain areas of the Himalayas.

A well thought out planning and management of these two resources which are the products of environments (for example, protection, conservation, augmentation of natural forests) ensures continuous supplies of water and energy. Environmental conservation is a sure way to put water cycle in order. In long term, environmental conservation achievable through ecological development (protection, conservation and enhancement of biodiversity in their natural habitats), would empower women folk for taking up new initiatives and implement ambitious plans in fragile mountain areas.

Water-energy-women issues do not find a centre-stage in the Millennium Development Goals set out by the United Nations in its Millennium Summit 2000. However, the issues relating to these aspects of mountain life are of critical importance and, if resolved, would contribute to achieve the Millennium Development Goals.

References

Bandyopadhyay, J. 2006. Integrated water systems management in South Asia: A framework for research. *CDEP Occasional Paper 09*. pp. 40.

Gurung, D.R., Bajracharya, S.R., Immerzeel, W. and Kulkarni, A.V. 2011. Understanding the dynamics of snow and glaciers in the Hindu Kush-Himalayan Region. *Sustainable Mountain Development*, No. 60. 17-21.

Sharma, B. and Banskota, K. 2006. Women, water, energy and the Millennium Development Goals: Lessons learned and implications for policy. *ICIMOD Newsletter*, No. 49. 3-8.

Sharma, B. and Giri, M. 2006. Women as energy entrepreneurs: Success from a pilot project in Phobjikha and Limukha Villages, Bhutan. *ICIMOD Newsletter*, No. 49. 21-22.

Singh, V. 1998. *Draught Animal Power and Sustainable Mountain Agriculture: A Study of Perspectives and Issues in Central Himalayas, India*. ICIMOD-MFS Paper No. 2.

Chapter 22

Industrial Use of Wastewater to Minimize its Misuse

Archita Rai, Neeraj Kumar and Sanchita Sharma

Global warming is a reason for the accelerated melting of glacier which make a reason for flood in the sea side area where lot of population lives. Any factor of generation of global warming affects the environment directly, so there is a need to minimize every cause to generate it. In this, the decrement of water level affects the people in two ways: one is the water access problem and second is the global warming which mainly occurs due to wastage and drainage of water. In this paper, there is a solution proposed to minimize the wastage of water through its industrial use. This paper also outlines about the fruitful use of wastewater but here we have not considered the cost aspect for the solution to minimize the wastage of water because lives are more costly than any cost at any given scenario.

Introduction

In today's scenario the water level of earth is a huge problem because day by day the water level is going down and as regards the present scenario, which shows that a huge amount of water type, including the rainwater and the water drained by domestic use, is going wasted which is further going to oceans and seas through the rivers. Here another thing is also to be summarized—that this drainage of water is an important cause behind the Global Warming. In this paper, we are proposing the economic model for the industrial uses of the wastewater for the fast developing nations like India where, no doubt, water is assumed as an economic good for any country as a limited resource mainly in case with developing countries which is indispensable (Perry *et al.*, 1997). This proposal is very fruitful for the countries having water problem in today's scenario or might be having in the near future, like in India or Pakistan. According to United Nations, 1.1 billion people are without sufficient water. By year 2025, 83% of rural people will be living in developing countries, where only 20% of the population will be having access to sufficient water supply (Rosemann, 2005). In this paper, the section one provides Introduction where literature survey and problems are defined with the outline about paper, section two describes the problem statement and existing solution for the problem, section three tells about some related work, section four provides the drawbacks and current results about the existing approaches, section five provides proposed solution and second last section six gives analysis after applying proposed solution and in section seven we suggest the conclusion with merits and demerits about the proposed solution.

Problem Statement

The problem tried to solve in this paper is about the industrial use of wastewater whenever it is of grey or black color, to solve the management problem of wastewater or make the fruitful use of wastewater. In scenario with developing countries like Pakistan where water availability has been declining at an alarming rate, it has shown that availability has less than 700 cubic meters per capita by 2025 (WWF-Pakistan, 2007). So after seeing these challenges, this paper tries to minimize the wastage of water to solve more water related problems.

Related Work

Before proposing a model to minimize the wastage of water, some tasks are important to accomplish which are:

1. Identify the authorize connections
2. Install water meters for the pipeline and give only necessary water into the supply lines
3. Change the faulty pipelines from the distribution
4. Minimize the carryout leak detection

Now after the minimization of misuse of good water there are some mechanisms for water treatment that can be used at multi story buildings, where the sewage should be treated in the basement itself and reused as make up water in the building's air-conditioning system. Chennai and Mumbai have been using treated sewage for various non-potable purposes. In Chandigarh, about 45 MLD of sewage is given tertiary treatment and then used for horticulture, watering of lawns etc. In Chennai, it is contemplated to treat 100 MLD up to tertiary level and use the same in major industries.

Water Treatment Technology

The basic objective of water treatment is the production of drinking water and water that is appealing to the consumer. Ideally, appealing water is one that consists of features of being clear and colorless, pleasant to taste, odorless, and cool. It is non-staining, neither corrosive nor scale forming, and reasonably soft. The next objective of water treatment is that it may be accomplished using facilities with reasonable capital and operating costs.

Surface Water Treatment System

For large scale water treatment, the mechanism UASB (Khalil *et al.*, 2008) is useful.

A full scale UASB plant of 14 MLD was constructed at Mirzapur for treating the domestic wastewater (Draaijer *et al.*, 1992) The state of Haryana almost entirely opted for UASB technology where 10 out of the 11 large plants were based on this. A top view and sectional view of UASB reactor is shown in Figure 22.1. Simpler wastewater feed inlet system in the UASB reactors is adopted to take care of choking, operation and maintenance problems surfaced at 5 MLD plant. However, in the ten UASB STPs designed for YAP in Haryana and recently in other UASBs. The feeding boxes, effluent gutters, baffle plates and gas collection pipes are also constructed with FRP material. For feeding pipes, HDPE (High Density Polyethylene) pipes are being used to distribute the wastewater uniformly over the surface of the reactor. For sludge discharge, CI (Cast iron) pipe is being generally used.

The other treatment unit in a water treatment plant is aeration, where water is brought in contact with atmospheric air to fresh surface water and also oxidizes some of the compounds, if necessary.

Figure 22.1
***Source*: Khalil *et al.*, 2008**

The next unit is chemical addition or flash mixer where coagulant (mostly alum) is thoroughly mixed with raw water by way of which neutralization of charge of particles (coagulation) occurs. This water is then flocculated, *i.e.* bigger floc formation is encouraged which enhances settlement. The flocculated water is then taken to sedimentation tanks/ clarifiers for removal of flocs and from there to filters where remaining turbidity is removed. The filtered water is then disinfected, mostly with chlorine and then stored in clear water reservoirs from where it is taken to water distribution system. However, before the supply of it for any purpose, the disinfection of water is important so it should be disinfected.

Proposed Idea

We have proposed here a model for minimizing the wastage of water and reuse the wastewater for the industrial use with the other uses. Here the basic idea behind the proposal is that, after doing a treatment of wastewater of a city, there should be a large reservoir between two or more cities on a definite height which is able to generate a perfect flow and this water should come back to the reservoir if necessary as per the figure given in Figure 22.2.

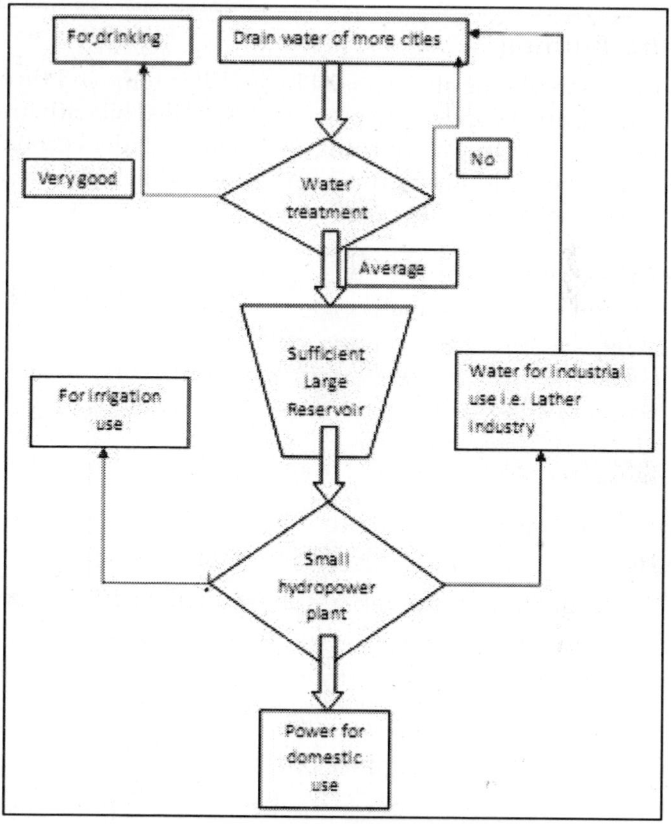

Figure 22.2: Water Flow in Industrial Use

In this proposal, power will be generated by a motor and remaining water may drain into the river as per Figure 22.3 described in tutorial.

Results

This methodology provides an economically good mechanism for the power generation for small area but when this technology will be ready to be applied in different parts of the country, it will be very good, hygienic and beneficial for the environment.

Conclusion

This technology is very beneficial for the developing countries

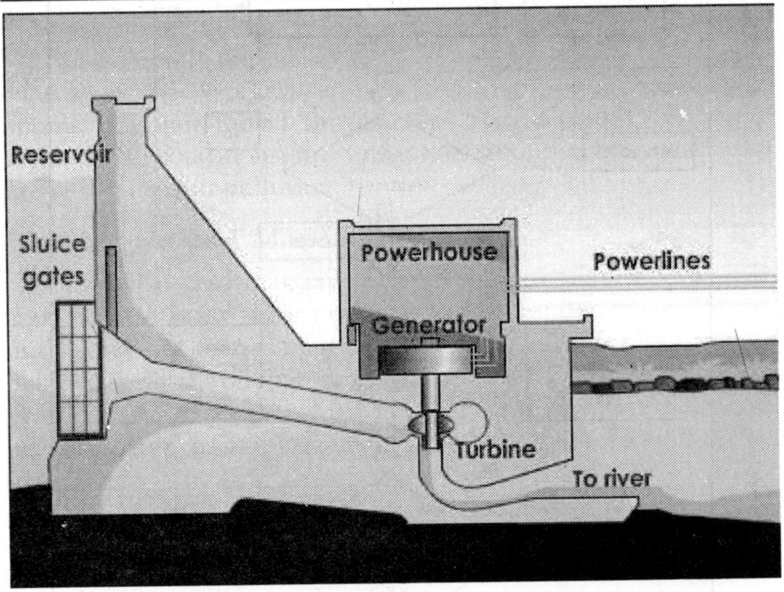

Figure 22. 3: A Schematic View of a Hydropower Plant
***Source*: http://www.tutorvista.com/content/science/science-ii/sourcesenergy/hydro-electric-power.php**

and for the hot countries, which which are faced with problem of global warming, water wastage and by the energy problems.

References

Central Pollution Control Board. *Status of Water Treatment Plants in India*. http://cpcb.nic.in/upload/NewItems/NewItem_103_statusofwaterqualitypackage.pdf

Draaijer, H., Maas, J.A.W., Shaapman, J.E. and Khan A. 1992. Performance of the 5 MLD UASB Reactor for Sewage Treatment at Kanpur, India. *Wat. Sci. & Tech*. Vol. 25, No. 7, pp. 123-133.

http://www.tutorvista.com/content/sscience/science-ii/sourcesenergy/hydro-electric-power.php

Khalil, N. Sinhab, R. Raghava, A.K. and Mittal, A.K. 2008. UASB Technology for sewage treatment in India: Experience, Economic Evaluation and its Potential in Other Developing Countries. Twelfth International *Water Technology Conference,* IWTC12 2008, Alexandria, Egypt, 1411-1427.

Perry, C.J., Rock, Michael and Seckler, D. 1997. *Water as an Economic Good: A Solution or a Problem?* London: E & FN Spon.

Rosemann, N. 2005. *Drinking Water Crisis in Pakistan and the Issue of Bottled Water: The Case of Nestlé's 'Pure Life'*. Swiss Coalition of Development Organizations, 2005.

WWF-Pakistan. 2007. *Pakistan's Waters at Risk: Water and Health related Issues in Pakistan and Key Recommendations*. Lahore: WWF.